THE PHILOSOPHY OF SCIENCE
APPLIED TO
MODERN PHYSICS

By Douglas E. Reinhardt
Ph.D., UNC-Chapel Hill

Key Subjects

An Alternate Physics, A Skeptical Approach to Modern Physics, Philosophy of Physics, Epistemology, Anti-relativity, Quantum Mechanics, Quantum Mysticism, Scientific Realism, Metaphysics, Cosmology, Critical Thinking, Metacognition, Research Methods, Logic, Rational-Empiricism, Logical Positivism

Key Question:
Does modern physics use a different kind of logic or thought process from that which is used in other sciences?

Copyright©2018: First Edition

by Douglas Reinhardt
All Rights Reserved

All images for cover courtesy of Wikimedia Commons.

Dedication and Acknowledgements

I dedicate this controversial book by an author who has ventured far outside his field to my loving wife, Peggy, who has been so tolerant and understanding of my need to make a statement on my perception of the "state of science" particularly in the field of physics. In our post-retirement years when folks are supposed to be sitting in rocking chairs, traveling and socializing with friends and family, she has supported me in every way despite my taking time away from her to pursue this passion.

I am also indebted to my science-oriented friends who have continued to stimulate my interest in hard science over the years. Those who argued for the standard theories of physics and resisted my classical tendencies have made me sharpen and fine tune my arguments for an alternate way of viewing the "nature of nature" and bring the "physical" back into physics.

And, to all my family, who may have felt neglected because of my solitary pursuits in reading, thinking and writing, I appreciate your support and understanding.

The Philosophy of Science

Table of Contents - toc

Abstract 5

PROLOGUE 6
The Outsider Perspective
Principles for Evaluating Modern Physics
Applying my Principles in a Debate with a Physics Student
So When can a Philosopher or Layman Have a Legitimate Opinion in Physics?

CHAPTER 1: LANGUAGE, LOGIC AND THEORY 30
Can Language actually Make a Difference in the Accuracy of a Theory?
Language and Mysticism
Dualistic and Monistic Philosophy in Physics
Critique of Bohm and Capra
Capra's Response to Criticism by Fellow Physicists
Piaget, Mysticism and Pre-logical Stage

CHAPTER 2: THE ROLE OF CONSCIOUSNESS IN PHYSICS THEORY 48
The Anthropic Principle
The Parapsychology and Physics Connection
Psychokinesis (PK) and Modern Physics
Near Death Experiences (NDEs), Soul Travel and Modern Physics
Consciousness: Another Well-Kept Secret
Holographic Universe, Consciousness and Information Theory
Field Theory, Information and Consciousness
Morphic Resonance and Field Theory
Can Materialist Physicists Rescue Quantum Weirdness From Mysticism and Spiritualism?

CHAPTER 3: PHILOSOPHY OF SCIENCE AND MODERN PHYSICS 74
Is the Connection Between Physics and Philosophy Dead?
What is Reality and How do we Know it?
Scientific Realism
Common Sense, Empiricism and Modern Physics
Demystifying Uncertainty: Particles, People and Probability

CHAPTER 4: THE SCIENTIFIC METHOD AND LANGUAGE 90
Epistemology and Ways of Seeking Truth
The Semantic Domains of Science: Physics and Metaphysics
Metaphysical vs. Mystical
Steps in the Deductive Method of Science
Metaphysical Rationalism in Modern Physics
Mathematical Proofs vs. Scientific Proofs

CHAPTER 5: CLASSICAL PHYSICS 113
Centrifugal Force: Fiction or Fact?
Centripetal, Centrifugal and Tangential Forces Demonstrated

Inertia, Momentum, and Force
Inertia of Rest Distinguished from Inertial Motion
Philosophy of Classical Physics vs. Philosophy of Modern physics

SUMMARY 129

References

Abstract

This book is about the philosophy of science, particularly epistemological issues relating to modern physics. It is written from the point of view of a philosopher of science who pays close attention to language and how it influences our logic and ways of thinking. The thesis of the book is that faulty use of language has led to some untenable theories in physics, and this may be the reason that it is difficult to find empirical support for such theories as String Theory or Quantum Loop Gravity. Several physicists have recognized that the trend toward metaphysics and away from empiricism has reached its zenith in String Theory and is symptomatic of a larger problem in Modern Physics. The word "logic" is derived from the Greek word, "logos" (which means "word") so language is the vehicle of logic. The two major terms that are misused in the opinion of the author are "space" and "time". Space and time are metaphysical concepts and when treated as physical entities, fantastic concepts emerge, such as "wormholes" where space can be folded up like a sheet of paper to enable space travelers to theoretically burrow a hole through the fold to take a shortcut to distant places and times within the universe. The guiding precept in the book is that qualitative language must precede quantitative language in building a theory. Sometimes physicists miss the conceptual forest for all the mathematical trees. In a word, they are blinded by the math. If the conceptual premises based upon certain assumptions are wrong, no amount of math can fix the theory. Ptolemy's geocentric theory based upon epicycles in the orbits of the planets is an example of a theory that yielded accurate mathematical predictions but was wrong because of systematic error. It is the view of this author that some modern physics theories partake of the same kind of epicycle-type fixes as Ptolemy's geocentrism. Thus, this book is written from the perspective of a fairly-well informed consumer of popular physics, a reviewer of popular physics literature and some professional literature on the subject, an anthropologist with an interest in linguistics, and most of all, a philosopher of science who is an adherent to the scientific method. Reification and mathematical abstraction have largely taken the "physical" out of physics. I would like to see a return to physical mechanisms to replace metaphysical abstractions such as "spacetime."

If one can imagine a tree metaphorically representing a theory, the roots would represent the foundational premises of the theory, the trunk the central concept, and the branches would represent the extensions and ramifications of the theory. While physicists argue about which secondary branches of the tree are correct, I am challenging the root premises of some physics theories which give rise to the trunk, primary branches and secondary branches. The premises of theory are rooted in language, logic, and concepts made from these elements of thought. If the roots are not well-grounded, the entire tree will fall.

PROLOGUE

The trend toward metaphysical mathematics and away from empiricism is epitomized by String Theory. Although this trend has reached its peak in String theory, this movement is symptomatic of a larger problem in Modern Physics which has culminated in this unscientific theory. Here is what physicist Lee Smolin (2006) has to say about this departure from science in the classical sense. In his critique of String Theory, Smolin here indicates that the problem is pervasive in Modern Physics.

I believe there is something basic we are all missing, some wrong assumption we are all making. If this is so, then we need to isolate the wrong assumption and replace it with a new idea. What could this wrong assumption be? My guess is that it involves two things: the foundations of quantum mechanics and the nature of time. . . More and more, I have the feeling that quantum theory and general relativity are both deeply wrong about the nature of time. It is not enough to combine them. There is a deeper problem, perhaps going back to the origin of physics (p. 256).

In examining the scientific method as related to Modern Physics, this book is about thinking. More specifically, it is about "thinking about thinking" or what has been called "critical thinking". As a student of science, I learned early on that science is mainly a thought process for discovering what is true in nature. The scientific thought process is called rational-empiricism or logical positivism by philosophers of science, and it is about demystifying what primal, pre-scientific humans had mystified because of their lack of understanding of nature. Furthermore, science assumes that the material world is real, not an illusion, and that our senses (and instruments that extend our senses) can give us a pretty accurate picture of nature - certainly accurate enough to ensure our survival as a species for the last several hundred thousand years.

The preceding paragraph is what I thought about scientific philosophy and the scientific method - until I encountered modern physics. While classical physics seemed like the other sciences I had studied with rational theories supported by observation and experiment, modern physics, including relativity and quantum mechanics, seemed like a return to mystical, pre-scientific thinking similar to the primal thinking I had studied in anthropology. Intrigued, I sent out a survey to physics professors asking if "neophysics" requires a different kind of logic or thought mode than other sciences – perhaps a kind of paralogic or even higher logic that I did not understand. Three physicists answered my survey, and two seemed offended by the term I had coined – "neophysics." All of them told me that "modern physics" (not neophysics) is the proper term, and that the answer to my question was an emphatic "no." No, modern physics does not require a different kind of logic – it uses the same logic as all the other sciences. However, Dr. Leonard Susskind (2006), Stanford Professor of Physics, clearly indicates that Modern Physics requires a different logic and thought process from Classical Physics. He says in a class lecture uploaded to YouTube:

All of Modern Physics has to do with those things which are beyond the intuitions we were able to get from the ordinary world…(Modern Physics) has to do with a range of parameters that humans or animals never experienced…(Physicists) had to invent new mathematics that was abstract (because) they couldn't visualize it…they (Physicists) learn to rewire themselves to develop intuitions to be able to deal with these new ranges of parameters…evolution simply didn't provide you with the means to visualize (quantum or relative phenomena)(Quantum Entanglement, Part I).

Then, I discovered Fritjof Capra. Capra also confirmed what I had been thinking about the "logic", or perhaps paralogic, of modern physics. In his fascinating book, *The Tao of Physics*, Capra said that modern physics partakes of the same kind of thought process as Eastern mysticism. In college, I had taken two semesters of Indian history with a heavy dose of Hinduism, and I found that Capra's stimulating book resonated with my own perceptions of modern physics. I, too, had observed the parallels between "neophysics" (Capra called it the new physics) and Eastern mysticism. The strange tricks that time plays with our perceptions in relativity seemed eerily similar to Hindu (and primal) concepts of time. And, apparent contradictions coexisting in the same quantum theory seemed strangely familiar to Hindu, non-categorical thought where contradictions are transcended. To wit, the claim of some quantum physicists that a cat can be dead and alive at the same time, if its life depends on a quantum probability, closely resembles the Hindu saying: "I exist and non-exist at the same time". Both concepts seem to use the same kind of mysticism even though quantum predictions may have been derived from sophisticated mathematics and experiments.

Furthermore, I was intrigued by how physicists use language. My studies in anthropological linguistics had taught me that the way a people perceive the world largely depends on their language, deeply embedded in their culture. For example, the Navaho language does not distinguish between blue and green. That lack of distinction does not mean that Navahos are color blind, but it does mean that they are less sensitive to these color categories and thus will make more errors than English speakers in tests requiring them to separate blue and green objects. In anthropology, the way a culture divides the continuum of reality into categories is called ethnoscience. While Fritjof Capra and David Bohm contend, true to Eastern mysticism, that the unity of reality cannot be divided into categories without distorting it, it is difficult to speak about any topic without the distinctions made possible by our categories of speech. The digit span of our brains is limited to 5-7 bits of information that can be held at once in our RAM – short term memory. In applying ethnoscience to physics, this book will examine the way physicists use the categories of space and time in theory building and how lack of careful definition of these terms has led to some rather other-worldly theories in physics. If space is void or vacuum, then how can physicists speak of space as being curved or folded up like a sheet of paper through which a wormhole can be burrowed to provide a shortcut to distant parts of the galaxy or universe?

I have been encouraged to find that some people inside the physics world have come to conclusions similar to mine as an outsider. Howard Hayden, retired physicist from the University of Connecticut, is one of those physicists who has dared to espouse heretical views on relativity. In reading his work, I have found support from the fact that someone inside the physics community has questioned relativity on the same basis that I have. Hayden makes it clear that he does not believe in the constancy of the speed of light in all circumstances and that this postulate of light speed constancy in special relativity is what has unnecessarily complicated general relativity in attempts to harmonize the two theories which are inherently incompatible. Furthermore, the work of Tom Bethell in conjunction with Petr Beckmann (summed up in *Questioning Einstein*) and William C. Mitchell's excellent book entitled, *Bye Bye Big Bang, Hello Reality*, let me know that I was not the only one who questioned whether relativity and quantum mechanics follow the scientific method. All these authors have the same problem that I do with the definition of space and time as material entities like matter and energy. These foundational concepts of space, time and spacetime in relativity and quantum theory have led to many erroneous conclusions in the opinion of the author of this book and the authors just mentioned. Furthermore, I found that physicists Smolin, Penrose and Woit have essentially the same problems that I have with String Theory, namely, that it is pure mathematical metaphysics and unscientific. Woit's critique of String Theory is embedded in the title of his book, *Not Even Wrong*. These critiques by physicists will be discussed later.

Every aggregate of people that interact frequently over time will develop a language, culture and a self-conscious group or society. This process is as inevitable as gravity causing matter to aggregate to form spheroid bodies like we see throughout the visible universe. Physicists are no exception to this law of social gravity. Physicists have formed their own culture with a special language including a mathematical language. As a group grows larger, it will begin to fission into subcultures, each with its own dialectic version of the language. Thus, we see the fissioning of physics into classical, relativity and quantum tribes, and the languages seem to contradict each other on certain key points. To wit, relativists tell us that nothing can go faster than light, but quantum physicists tell us that particles that are entangled can convey information instantly (faster than light, of course) over vast distances.

And so, I write this book in a very personal style, freely using the first-person pronouns (I, me, my) because it is a personal odyssey to find the truth about nature and the universe. I come to physics on my own, and rather late in life, because I feel that physics deals with the most fundamental questions of philosophy: "What do we know, and how do we know it?" In the more sophisticated language of philosophy, this quest is called **epistemology**. Since it is epistemological in nature, this work is not meant to be a polemic against physicists, although it may seem so. It is a call for physicists to give more attention to their thought processes and language in building theories that will stand the test of time. It is thus an invitation to meta-cognition, thinking about thinking, critical thinking, and ultimately, it is about the philosophy of science itself. It asks the fundamental question of "What is the scientific method, and does modern physics deviate somewhat from that method?" Furthermore, is the scientific method changing to accommodate a broader range of thought processes that are mystical in nature? Has the gap been narrowed between science and mysticism (and perhaps religion)?

This book is a personal odyssey in the sense that I, at first, accepted the strange theories of relativity and quantum mechanics on blind faith, trusting in the authority of famous physicists such as Carl Sagan. Sagan was my hero, and I accepted his teachings with the awe and credulity of a religious devotee. However, our college experience teaches us to be "critical thinkers", i.e., to examine ideas and theories for ourselves rather than accepting authority. While critical thinking about subjects outside one's field may be seen as hubris by the experts in that field, it is never a good idea to accept authority blindly -- and when experts disagree on basic philosophical issues that transcend any given field of science, then the **philosopher of science** can have a legitimate opinion in my view.

Accordingly, as I began to study physics for myself, I became more and more skeptical of the theories that were being passed off as the "laws of nature". It seemed that some inferences drawn from the data went far beyond what the facts would allow. For example, how can one reasonably infer from the uncertainty of where a particle might be in a quantum experiment that the whole universe splits into many universes so that every statistical possibility of the particle's location is realized in some universe. This "many worlds theory" would imply that if my winning the lottery depends on some quantum event (such numbers being generated by random emission of radioactive particles), then in some universe, I have won the lottery -- along with millions of other people who have won the lottery in their respective universes. I, along with millions of others, just wish to win the lottery in this universe. Such thinking truly requires a **quantum leap** from a subatomic particle to the whole universe or multi-verse. Carl Sagan said it well: "Extraordinary claims require extraordinary proof." Of course, he was referring to psychic phenomena, but I think the same principle applies to physics as well.

In considering such theories as the many worlds theory, I became even more fascinated with the way physicists use language to describe physical and metaphysical phenomena. Language, after all, is the basis of logic as stated previously. Thus, a theory begins in language and logic. If the language is not accurate, then no amount of mathematical manipulation can make it right. Therefore, I believe that the theories of physics can be debated at the conceptual level without reference to mathematics, since the math can be no better than the verbal concepts and assumptions it is predicated upon. For example, if string theory's conceptualization of fundamental matter as tiny, one-dimensional strings that vibrate in 11 dimensions is incorrect, then all that complex calculus supporting string theory is not describing any physical phenomena – even though it may produce some pure mathematics. In the same manner, analysis of the game of chess can generate pure mathematics, and a computer can be programmed using this math to defeat the world chess champion. However, chess mathematics does not describe any physical system so far as I know. This math is purely metaphysical in nature and belongs in the universe of pure mathematics – not physics.

Physics is sometimes called the "Queen of Science" and is considered one of the hard sciences, as opposed to soft sciences such as psychology and the social sciences. Ironically, the soft sciences are becoming more hard-nosed empirical and materialistic while physics and some elements of biology are becoming more non-materialistic and informational. In the psychology courses I took in college, there was little talk of "the mind" which is regarded as a spooky concept and unobservable. Psychologists seem to prefer to talk about the brain and behavior which can be observed and measured. Sociology and anthropology have moved in the direction of biological reductionism and away from the "nurture" part of the equation.

As indicated, some biologists are also moving in the direction of the mystical and metaphysical. These prominent biologists are rejecting the Darwinian mechanism of evolution and are espousing some form of intelligent design to account for "irreducible complexity" and the rapid spurts in evolution known as punctuated equilibrium. The intelligent design paradigm in biology is very much akin to the *strong anthropic principle* in physics in my estimation. The irony of ironies is that physics is becoming less physical while the soft sciences are becoming more physical.

Despite my varied academic background, many physicists will see me as arrogant and a shard of cracked pottery for writing a book questioning the language of physics, but perhaps the book will offer a fresh perspective on the thought processes of physics as it relates to the philosophy of science. Perhaps someone trained in physics would find it difficult to break through the entrenched thought molds to see another perspective. We are all familiar with the tendency for "groupthink" that is widespread, if not universal, in social interaction. Conceivably, this book could lead to some amendments to theories which I believe are characterized by hyperbole, unjustified inferences and imprecise definitions. Maybe it will also appeal to fellow philosophers in hot pursuit of the truth, who have had a similar struggle in buying the mystical logic of relativity and quantum physics. In this bold venture into another field of knowledge, I speak only for myself. I do not pretend to speak for other philosophers who may have little interest in this subject, and I certainly do not speak for physicists. My aim is to speak as a **philosopher of science** who is searching for truth through the scientific method which has served us pretty well in unlocking some of nature's secrets in the past few centuries. In my view, some work in modern physics deviates from the principles of the scientific method.

To put my prologue in a nutshell, this book is about examining the language and logic of modern physics and an evaluation of whether the scientific method founded upon Western logic is being followed in formulating and testing theories. The book is also an extension of Capra's work showing the close connection between Eastern mysticism and modern physics. I also see a connection between the thought processes of primal peoples known to anthropology and modern physics. I believe the similarity in thought in the face of mystery is an innate response based upon brain structure. Robert Ornstein has identified Western linear logic with the left hemisphere of the brain which is the seat of language. Contrastingly, he identifies Eastern, non-linear, holistic thinking with the right hemisphere. Robert Ornstein's thesis along with Piaget's study of the stages of cognitive development in children will be used to analyze how brain structure and function impact theory formation. Since this writing is a personal quest, perhaps I will undergo another metamorphosis and return to a more spiritual, mystical way of thinking that characterized my youth. Perhaps I will find an Hegelian syntheses between East and West, mysticism and rationalism, left and right hemispheric thinking. After all, it must have been adaptive to have two ways of thinking in human evolution and in the development of a complex brain. It is with humility that I recognize that the universe is bigger than my philosophy and that there may be a multiverse with multiple dimensions which may be holographic in nature. If so, I want to know it. If these questions intrigue you, as they do me, then read on.

Why would anyone be interested in what a Philosopher of Science thinks about modern physics?
Someone said it well when s/he stated that all of us are "addicted to our beliefs." Physicists are like the rest of us humans in that regard, I believe. The ideal of keeping an open mind in science is just that – an ideal that we humans fall short of. When our theories, to which we have an emotional attachment, are challenged, we are almost instinctively motivated to defend them in the same manner that we would defend our territory – especially when those challenging ideas come from outside our territory from another field. When forced by logic and evidence to abandon our addictive beliefs, we suffer from a kind of "withdrawal" similar to withdrawal from a drug. I know this "withdrawal" from personal experience after having to abandon cherished beliefs in my college experience in which my traditional ideas were challenged. So, to break through addictive beliefs sometimes requires an outside perspective. Now, perhaps most ideas coming from outside a well-established field will be nonsensical, but occasionally, good insights will emerge. Are my

ideas nonsense or of some value? I leave the answer to that question to my readers. All I ask is that you give a fair and objective hearing to my outsider perspective.

The Outsider Perspective

The theories of relativity and quantum mechanics have considerable momentum in physics which makes it difficult to overturn or even modify them. After theories have been well-established as in Modern Physics, most new theories hatched by mavericks outside the field will be considered cracked pottery at best, but there is a possibility that a few of these theories will be solid and revolutionary. It is rather like mutations in genetics. After the human organism has been honed by evolution over millions of years (including non-human ancestors), most mutations will be harmful or perhaps have no effect at all, but in rare instances, there will be a mutation that is helpful to the survival and propagation of the species. Likewise, in society, most deviance is harmful (such as crime), but, in rare instances, some forms of deviance can lead to positive changes in the social fabric. Remember that Martin Luther King was a deviant who violated many laws that he believed were unjust. So it is with science. Scientists are not immune to the social forces of conformity, groupthink and mental sets that prevent them from seeing alternatives and fresh perspectives on physical reality.

It could be debated as to whether Einstein was a maverick when he introduced the Theory of Special Relativity while working as a clerk in a patent office outside the walls of academia. While he was not connected to a university, he had finished a degree in physics and was in touch with fellow students of physics while working on his monumental paper on the *Electrodynamics of Moving Bodies*. So we could say he was an in-between maverick – not officially connected to an academic institution but not out of touch with those who were connected. Certainly he was not one of the herd and often cut classes in order to study on his own. Cox and Forshaw (2010) indicate that

An unfortunate consequence of Einstein's apparent isolation from the mainstream is the modern temptation to look upon him as a maverick who took on the scientific establishment and won unfortunate because it provides inspiration to any number of crackpots who think they have single-handedly discovered a new theory of the universe and cannot understand why nobody will listen to them. In fact, Einstein was reasonably well connected to the scientific establishment, although it is true that he did not have an easy beginning to his academic career (Kindle Book 15%).

As indicated above, most novel ideas coming from outside academia will be like useless mutations in biology, but it is my belief that a few innovative ideas can break the mental molds (that those trained in physics find it difficult to break through) and can offer some fresh perspectives to the field. Certainly, the early physicists had few if any peers with whom to discuss revolutionary ideas. With whom did Galileo and Newton speak about their ideas, and were their writings peer reviewed? How was Michael Faraday, who had little formal education, able to discover the principles underlying electromagnetic induction, diamagnetism and electrolysis? Much of his work was later given mathematical and formal description by the more-educated James Clerk Maxwell. In another field, Gregor Mendel worked as a lone researcher in his pea patch in a monastery, without peer interaction or peer review, to discover the laws of genetics.

However, one might argue that the founding fathers of modern physics could act as mavericks when physics theory was more basic in its infancy, but now physics has grown to become a more complex and mathematical discipline so that those uninitiated into its secrets could hardly understand, much less make a contribution. While there is much truth to this argument in regard to specifics of theories and technology, I still believe that going back to the basics and analyzing the language, logic and the premises on which theories are constructed might be better performed by a philosopher of science outside the field who has not been indoctrinated into the discipline. According to many in the field, physics students are not taught critical thinking as they are in other disciplines. Instead instruction tends to be more authoritarian. Here's what Essen has to say about such training: "students are told that the theory (relativity) must be accepted although they cannot expect to understand it. They are encouraged right at the beginning of their careers to forsake science for dogma (Kelly 2005: p. 285)."

Tom Bethel echoes this observation:

Where Einstein's relativity is concerned, they (physicists) have been trained to disregard all dissenting views. Only the strictest orthodoxy is allowed...the theory's premise (is not) questioned. If new recruits to academy pursue an academic career, they won't be open to any dissent or doubt about special relativity. Inconsistent experimental results will somehow lie outside their field of vision (2009: p. iii).

This authoritarian attitude regarding theory is foreign to me, coming from the field of social science where critical thinking about established theory is encouraged and even required by professors of the various disciplines. Students are invited to critically evaluate theories analyzing their strong points and weak points to form their own theory from a synthesis of ideas. Social scientists recognized that no theory can tell the "whole truth" because our limited attention span as humans allows us to consider only a small number of variables at a time while nature operates on many variables interacting simultaneously. Admittedly, social science is more fluid and less amenable to precise mathematics because the phenomena dealt with are perhaps the most complex phenomena in the universe - the human mind and the interaction of these minds in the social realm.

If nothing else, this book should provide significant feedback to the writers of popular physics to prompt them to justify their arguments regarding seemingly outrageous theories, rather than papering them over with the cliché – modern physics is "counter-intuitive". If it is too difficult for the layman to understand, then why try to popularize it? Fortunately, the internet provides a medium for the so-called mavericks in physics to debate controversial theories and offer varying perspectives.

What this book may have to offer might fall into the category of the philosophy of science as it relates to language. More specifically it aims to show how language itself tends to shape our perception of the world. Special language (including mathematical language) develops in every group including scientists, and it definitely shapes the way scientists see the world. The two examples of this tendency as shown numerous times in this book are the terms "space and time". The way physicists have defined space and time has shaped their perception and theories of the cosmos. The reification of these abstract terms has led to some erroneous conclusions about the universe, in this author's opinion.

Because of the human tendency to build mental sets and engage in groupthink, sometimes it is useful for someone outside a field to bring a fresh perspective to break through those thought molds. A glaring example of the dangers of groupthink is the Dalkon Shield fiasco. The Dalkon shield was a contraceptive intrauterine device that was marketed by the A.H. Robbins Corp. At a meeting to make the final decision on whether to take it to market, one of the scientist who had some serious reservations about the device was shut out of the meeting, and the rest is history. So many women suffered severe physical ailments such as serious pelvic infections leading to infertility or even death that the A.H. Robbins Company had to file for bankruptcy due to millions paid out in law suits (Kolata 1987). I'm not suggesting that groupthink about theoretical matters in physics could pose physical dangers to anyone, but it could pose dangers to developing sound theories that will stand the test of time.

Furthermore, familiarity with a thought process tends to desensitize us and turn off critical thinking. Then when an outrageous theory is proposed, it does not seem unreasonable to the insider. For example, a theory that suggests that the universe and its history were created by human consciousness going back in time may not sound outlandish to the insider because of familiarity with such ideas, but to the outsider, it arouses serious doubts. Or a theory that suggests that for every quantum probability, a new universe is spawned so that probability can be realized somewhere would seem to be beyond prove-ability and therefore unscientific, but to someone familiar with quantum theory, it might seem quite normal. However, as Sagan said: "Extraordinary claims requires extraordinary proof," and I would think that the many worlds theory would require some extraordinary proof.

In reading many books on relativity by well-known physicists, I have read few authors who question the fundamental language of relativity, especially such key concepts as space and time. With metronomic regularity, the great majority accept space and time as physical players in the interactions with matter and energy. Few that I have ever read say that perhaps space and time are abstractions or metaphysical notions that provide the conceptual framework for the interactions of matter and energy. Most of them will say that Newton conceived the universe with space and time playing passive roles in the background, independent of the interactions of mass and energy, but then popularizers of physics will proceed to repeat hypnotically that Newton's conception of the universe has been clearly superseded and proven wrong by Einstein's relativity.

Physics is certainly a specialized field that requires a tremendous amount of mathematical knowledge, and some physicists feel that it is an encroachment on their territory when someone outside their field raises a question about a theory. However, when physicists leave the realm of the empirical and speculate on the nature of reality and consciousness, then they have entered the realm of metaphysics and philosophy where philosophers and laymen can have an opinion – especially if there is disagreement among physicists on a theory. When a physics theory is based mostly on mathematical assumptions rather than observation, then an informed person can reasonably challenge such a theory. This is the guiding philosophy of this book – that all science begins with language and, if the language is wrong, then the theory is inaccurate, and no amount of math can make it right.

The key critique of physics is that it has taken the "physical" out of physics and replaced it with mathematical metaphysics and abstractions with little basis in empiricism. My recommendation, for whatever it is worth, is to get back to the basics and put physical forces back into physical theory – as classical as that may sound. Here are the principles I shall follow in evaluation scientific theories.

Principles for Evaluating Theories of Modern Physics

Modern Physics is an example of what happens when mathematical metaphysics is not disciplined by empiricism.

To use a couple of analogies - theory is like blinders on a horse that help prevent spooking or like a filter which allows some frequencies of light to pass while blocking others. Thus, theory opens our eyes to certain aspects of reality and blinds us to other aspects. The ideal theory is one based on the assumption of rational-empiricism and takes into account all relevant data and is not contradicted by any of the data. As the oath taken in a court of law says: "I swear to tell the truth, the whole truth and nothing but the truth…" so an ideal theory should aspire to tell the whole truth. Theories are usually incomplete because they do not take into account all the variables that cause a particular effect. The limited digit span of our brains makes it difficult to tell the whole truth. My analysis of the scientific method and theory formation as it relates to modern physics shall involve these principles.

1) Technology does not necessarily prove any particular theory. We often hear it said that if a particular theory were not true, a certain technology would not be possible. The problem with this idea is that there are usually several competing theories to explain even basic technology such as centrifuges. Despite the name *centrifuge*, some physicists deny the existence of "*centrifugal force*" and say there is only centripetal force operating in curved motion. Others say that two forces are needed to create curved motion: centripetal and tangential force. Still others argue for centrifugal force. Nonetheless, centrifuges continue to work regardless of which theory one espouses. Likewise, now some theorists indicate that the "Bernoulli principle" is not the reason airplanes fly despite the fact that scientists have thought this principle is the reason planes fly for over 100 years. Furthermore, electrical engineers often speak of a black box where things happen that make the system work and yet no one understands why it works. The black box function has been found by trial and error and serendipity - not theoretical prediction. Moreover, physicists still debate what causes a fluid to flow through a pipe. A modern example of the relationship between technology and theory is that no theoretical interpretation of quantum mechanics can claim that televisions, computers, particle colliders and other technologies based on subatomic particles or waves depends on their particular interpretation to work. To wit, does electronic technology work because of the Copenhagen interpretation, Many Worlds, EPR, or Bohmian Mechanics? And how does the uncertainty principle relate to the certainty that is needed in making cesium clocks and finely tuned electronic equipment work with such predictability and *certainty*?

John Gribben (1984) says that

Without the equations of quantum mechanics, physicists would be unable to design working nuclear power stations (or bombs), build lasers, or explain how the sun stays hot. Without quantum mechanics, chemistry would be in the dark ages, and there would be no science of molecular biology – no understanding of DNA, no genetic engineering – at all : Kindle Book 3%).

I can see how Pauli's exclusion principle would certainly help in determining the valences of elements and help predict the outcome of chemical reactions, but Gribbin must recall that the

Periodic Table was created in 1869, long before quantum theory was conceived. And electrical and electronic instruments were developed by Faraday even before Maxwell's equations described how electromagnetic forces work, and Maxwell's unification of electrical and magnetic forces (mid 1800's) predated quantum mechanics (1920's). Michael Faraday, who had only a basic education, in the early 1800's had invented electric motors, electrolysis, generators and a number of other things prior to Maxwell, relativity, and quantum mechanics which attempts to explain the behavior of electrons and other quantum particles. Furthermore, Edison with all his electric inventions confessed that he was no theorist. What we see is that theory often comes after technology, and then there are several competing theories as to how a particular technology works. This is not to say that theory is not helpful in developing technology, but there is certainly no one-to-one relationship between theory and technology in the way Gribbin has suggested. Generally speaking, technology is "how" something works, and theory is "why" something works.

It is interesting that the same John Gribbin who made the statement above also said:

We still don't know what electrons (or other quantum entities) are nor how they do the things they do…What makes it all the more remarkable that without knowing what quantum entities are or how they do the things they do, by knowing that they do do certain things when prodded in certain ways, physicists are able to use quantum entities. This is a bit like learning to drive a car by learning how to manipulate the controls, without having the faintest idea what is going on under the bonnet (2009: p. 14).

So, according to Gribben, without a theory to explain why quantum phenomena behave the way they do, scientists can still manipulate particles to produce technology that works consistently. Thus, he makes my point precisely, but in contradiction to his original point that certain technologies work because of certain theories. Hence without raising the hood on the atom, scientists can still manipulate its power to drive technology – just as one need never learn to raise the hood on an automobile to drive it.

2) A mathematical proof is not the same as an empirical proof.

A mathematical proof proves mathematics; it does not prove physics. A bad theory cannot be papered over with mathematics to make it right because math can be manipulated to make a theory consistent and thus seem right.

A mathematical proof such as a geometric proof is achieved with pure logic, not observation. A geometric proof is performed in the Greek way, that is, the proof cannot be accomplished with measurement because measurement is always inaccurate to some degree, to wit, measurement may be a close approximation but is never exact. A mathematical proof begins with axioms that are so obvious that no rational person can deny them and proceeds with step-by-step logic to show that more complex things are true. Hence if we have arrived at a proof that says that the area of a rectangle is length times width, then, if we draw a diagonal line connecting opposite angles, the two triangles formed are equal to each other and occupy ½ the area of the original rectangle. If one tries to measure this area, however, there will be a degree of inaccuracy.

On the other hand, a scientific proof is opposite from a mathematical proof. We may start with a logical statement or hypothesis, but the hypothesis must be measured against the empirical world because nature does not always conform to our logic. Hence Aristotle's logic of falling bodies (an object weighing 10 times more than another object will fall 10 times faster) was tested by Galileo and found to be in error. Aristotle was trying to use pure Greek logic to describe nature. *Therefore, logic proves mathematics, but empiricism proves science.* Much of modern physics seems to involve more mathematical than empirical proofs. String theory, according to Kaku, indicates that when dimensions exceed 12, the math shows that the universe becomes unstable (2011: Youtube video: Michio Kaku Explains String Theory). The question is: What does nature say about extra dimensions? Would nature agree that 13 is an unlucky number, or would nature say there are no dimensions other than the three we observe? Mathematics, like language is an approximation of reality: Physicist Paul Davies (2010) says that:

Many of my theoretical physicist colleagues do indeed regard ultimate reality as vested in the subset of mathematics that describes physical law…the logical conclusion…is to treat the physical universe as if it simply is mathematics (p. 67).

I would say that there may be mathematical regularities in nature, but human mathematics, like qualitative language, is an approximate representation of nature's mathematical laws which are much more complex than human mathematics. Take the value of pi which is said to be transcendental and incalculable. Yet, we know logically that the relationship between the diameter of a circle and its circumference is an exact geometric relationship and thus is not transcendental. Furthermore, our base-10 mathematics can only approximate the value of pi with millions of decimal places that are said to bring us closer and closer but never quite gets us to pi. Consider the fact that there are many physics equations that use pi to describe curvature. But because math can only approximate pi, these equations may come close to describing physical reality but never quite close the gap.

Consider the formula for calculating time from the oscillations of a pendulum clock.

$$T = 2\pi\sqrt{L/g}$$

Here, it is apparent that pi is a key component of the equation because the pendulum swing is an arc of an imaginary circle. However, pi is conceived as a never-ending decimal value that is transcendental; therefore, it introduces a slight inaccuracy into the pendulum formula since pi is an infinitely-repeating number that must be rounded off. Thus the law of the pendulum may be a perfect law, but our language and math are approximations of that law.

Similarly Davies (2010) says that the "laws of physics are normally cast as differential equations, which embed the concept of real numbers, and of infinite and infinitesimal quantities, as well as continuities of physical variables, such as those of space and time (Kindle Location 1527)." Again, we see that infinities, like the infinite decimal places of pi, can only be approximations of the finite phenomena that our math attempts to describe.

Michio Kaku quotes Einstein as saying: "Behind every great theory there is a simple physical picture that even lay people can understand…if a theory does not have a simple underlying picture, then the theory is probably worthless. The important thing is the physical picture; math is nothing but bookkeeping (Laureyssens 2009)."

Einstein is also quoted as saying: "I don't believe in mathematics…As far as the laws of mathematics refer to reality, they are not certain; and as far as they are certain, they do not refer to reality" (Albert Einstein Site Online).

I would simply add that mathematics, like qualitative language, is an approximation of reality – it is not reality itself any more than an image is the same thing as the object. It is the means and not the end in physics and all other sciences. William Mitchell (2002) said it well:

During this century, ever since Einstein came along, it has become fashionable for theorists to write equations and then, because it was possible to write them, decide that they must represent reality. They have apparently forgotten that mathematics is merely a tool that can be used as an aid in describing reality…Thomas Phipps has written…'The mathematical servant has become the master' (p. 240).

The Rayleigh-Jeans math underpinning the "ultraviolet catastrophe" is a perfect example of how mathematics may not accurately describe what it purports to describe. Quantum theory was born when Planck found a math and Einstein found a theory that would rescue physics from the ultraviolet catastrophe. Ironically, even though quantum theory got off to a good start with this development showing that empiricism trumps math, quantum theory has largely fallen into the thought trap that if the math works, the theory must be true. String theory is an excellent example of this thought trap.

Numerology is an ancient belief. The idea that the locus of mathematics is in the external world perhaps dates back to Pythagoras and his disciples who, some historians allege, believed that numbers are sacred and represent the ultimate reality. An apocryphal story alleges that Pythagoras drowned a student who discovered an irrational number because all numbers were conceived to be perfectly rational. Newton apparently believed that God is a mathematician who had founded the universe on mathematical laws that were discoverable by humans. The connection between math and religiosity is expressed by Edward Everett who said:

In the pure mathematics we contemplate absolute truths which existed in the divine mind before the morning stars sang together, and which will continue to exist there when the last of their radiant host shall have fallen from heaven (White 1956: p. 2349).

A more empirical view of mathematics is expressed by physicist P.W. Bridgman who averred that "it is the merest truism, evident at once to unsophisticated observation, that mathematics is a human invention" and Edward Kaner and James Newman added that "We have overcome the notion that mathematical truths have an existence independent and apart from our minds. It is even strange to us that such a notion could ever have existed (White 1956: p. 2349-2350)."

Obviously, these declarations of the locus of mathematics were made before quantum mechanics and string theory which rely heavily on the idea that mathematical truth is inherent in nature. Perhaps modern physicists do not associate math with specific religiosity, but it is often treated religiously.

3) A set of data does not necessarily support any one theory.

The same set of data can give rise to a number of theoretical interpretations to explain the same phenomena. As indicated above, quantum mechanics involves several competing theories to explain the interference patterns seen in double-slit experiments. To demonstrate how a particular theory may show consistency with the facts but be wrong in the mechanism, let us consider two theories regarding the incest taboo. All societies known to anthropology have some kind of incest taboo – although the group within which one cannot marry (or breed with) can vary from society to society. Many pre-literate societies explain this taboo by saying that marrying or having sex with a close relative makes the ancestral spirits angry; therefore, the spirits curse the child that comes from such unions with deformities and disease. It is a theory that works, but most modern people with a scientific, materialist mind-set might say the mechanism is wrong. Many people today would probably agree that close in-breeding makes it more likely that harmful, recessive genes will come together and cause abnormalities.

This "primal" incest taboo theory works pretty well as does Ptolemy's geocentric view of the universe and the theory of the ether. If enough epicycles are allowed, Ptolemy's theory accords very well with observation. Furthermore, there can be systematic errors in the theory that make it appear to be right when it is wrong, or there can be constants and correction factors (hedges) that make it appear to be right. Ptolemy's epicycles are good examples of "fudge factors" called correction factors added to a theory to produce systematic errors that make the theory appear to be consistent with celestial data and yield accurate predictions. As a result of its accuracy, Ptolemaic theory was accepted for some 1300 years as the way the universe works. "As an indication of exactly how good the Ptolemaic model is, modern planetariums are built using gears and motors that essentially reproduce the Ptolemaic model for the appearance of the sky as viewed from a stationary Earth (Iowa State Dept. of Physics and Astronomy 2001)." And, let's not forget "ether theory" which is incorporated into the much revered Maxwellian equations, and some physicists still hold to some form of this theory in the post Michelson-Morley world.

Likewise, when the Michelson-Gale experiment found that westbound light shows a higher relativistic speed that eastbound light, one headline read: "Michelson proves Einstein's theory." Another read: "Ether drift is confirmed. Rays Found to Travel at Different Speeds When sent in Opposite Directions (Bethel 2009: p. 122)."

Any time a patch, such as a constant, has to be inserted into a theory to make it work, a red flag of caution should be raised. Sometimes these amendments to a theory are necessary because reality can be subtle and complex. As Einstein supposedly said: "Subtle is the Lord". So how do we know when an epicycle-type patch to a theory reflects nature and when it is an *ad hoc* measure to force a theory to fit the empirical world? That is a tricky question and one which we shall explore in the pages ahead. In the thesis of this book, a linguistic analysis is a first approach to separating fudge factors from reality. Is the language used in an illogical way? Are concepts redefined to make things work?

I am going to be so bold as to assert my belief that some of modern physical theory resembles the Ptolemy illusion and will eventually be overturned. When there is great mystery surrounding a phenomenon such as quantum mechanics, the human tendency is to turn to metaphysics and mysticism. Later we shall discuss outright admission of mysticism in modern physics by Capra, Bohm, Heisenberg, and Bohr.

4) A theory that generates fairly accurate predictions is not necessarily correct in describing the mechanism as in the Ptolemaic fallacy.

The purpose of a hedge is to make a theory unfalsifiable. A hedge can sometimes be a constant or other fix to make a theory seem consistent with observation or experiment. Deductive bias in a theory can lead to hedges that enable a theory to be self-fulfilling.

Consistency is not always the same thing as accuracy. In the social sciences, we certainly find this to be true. In the early part of World War I, recruits were given IQ tests to determine what military occupation they should be placed into. The initial findings were that Whites on average scored about 20 points higher than Blacks. This finding fed a historically biased stereotype, namely, that Black people are not as intelligent as Whites due to inferior genetics. However, some analysts began to look for hidden variables in the data -- and they didn't find them on the other side of the universe -- they were found to be very local. When these analysts looked at the situation on a state-by-state basis, they found that that Blacks in some Northern states scored higher than Whites in some Southern states and that IQ scores were directly correlated with per capita spending on education in any given state. When all these factors were accounted for, the dramatic differences in IQ scores disappeared (Sowell 1995: 70-79). Hence, the hidden variable was right under the nose of those who believed in the heritability of IQ, but their bias prevented them from seeing it. Despite a faulty premise, the theory of racial superiority in intelligence predicted IQ scores for Whites and Blacks pretty accurately but attributed the wrong cause for the differences measured. Similarly, many times we hear that relativity, or some other theory is correct because it produces some accurate predictions.

As indicated above, Ptolemy's geocentric theory, with its many epicycles, accurately predicted the motion of the planets despite the fact that Ptolemy got things backwards as to which body was orbiting the other. Some planetariums still use the Ptolemy model to project the motion of planets and stars on their domes. Hence Ptolemy's geocentrism made accurate predictions but the mechanism was wrong. However, the predictions it made were, for all intents and purposes, as accurate as Copernicus's heliocentric theory. And, in Einstein's relativity, the view of an observer on the earth would be as valid as the view of an observer on the sun – thus supporting the geocentric theory from earth's perspective.

In the same way, one might argue that ether theory, an assumption in Maxwell's equations, produced some accurate predictions. Consider the following statement regarding prediction and validity.

The successes of the Lorentz transformation, the results of the Michelson-Morley and Ives-Stilwell experiments, and the numerous correct predictions guarantee the validity of the Einstein theory.

Peter Beckmann *(1987) responded to the foregoing statement this way*:

A thousand confirmations of a theory do not prove it, for a single discrepancy can destroy it - as shown by the ether theory, which also boasted an uncommon number of correct predictions in its day. Moreover, certain aspects of a theory do not get verified until challenged by a rival theory. (As an historical curiosity, one might add that the three authors above, Hendrik Lorentz, Albert Michelson and Herbert Ives did not accept the Einstein theory, and remained resolutely opposed to it to their deaths in 1927, 1931 and 1953, respectively).

A List of Ptolemaic, Epicycle-type Hedges in Physics Theory

a>**Ether** as mentioned above

b>**Massless Particles:** Photons are said to be massless particles even though they have energy and energy has mass according to $E=MC^2$. However, relativity indicates that any mass that accelerates to the speed of light will gain infinite mass. Since photons don't have infinite mass, they must be massless – circular logic.

c>**Nature of Space:** Special Relativity indicates that space is vacuum (nothing) and General Relativity indicates that space is a medium that produces gravity when warped. One possible hedge comes from quantum theory which indicates that space is something and nothing, so it can be nothing in Special Relativity and something in General relativity.

d>**Faster-than-Light Space Expansion:** When Big Bangers claimed that space is expanding faster than light and that is the reason galaxies are receding from each other in a red shift, the obvious objection is that nothing can go faster than light according to relativity. The hedge is that it is not the galaxies that are moving faster than light, it is space that is expanding faster-than-light carrying the galaxies with it while the galaxies are at rest in space, i.e., not moving at all. The ultimate hedge is that space is nothing, and nothing can go faster than light without violating the speed of light. Of course, space is treated as "something" when convenient.

e>**String theory** perhaps has more hedges than any theory because it is the most unscientific and needs much hedging. When string theory, which was supposed to produce one unifying equation, produced five equations instead, the hedge for that conundrum was to produce M-Theory and Superstring theory. The author of M-Theory said that the five string equations were manifestation of one equation. However, M-theory/Superstring theory has produced 10^{500} equations instead of one. The grand hedge is to say that the 10^{500} equations represent a multiverse with each equation defining a type of universe, so there is one unifying equation for each universe. However, we don't know which one describes our universe. So, we're united in our local universe, but we don't know which equation unites us.

d>**Many Worlds Interpretation rescues grandfather paradox**: Many argue that relativity makes possible time travel, but the inevitable question arises: "What if I go back in time and kill my grandfather before he fathers the son or daughter who is my parent?" Many Worlders come to the rescue and say that the universe splits, and in one universe I continue to exist and in the other, I don't exist. Of course, for the MWI to work, theoretically, the split has to depend on a quantum probability. Failing that, perhaps I would go out of existence.

e>**Decoherence hedges Many Worlds**: When asked why we can't see all these universes being formed in MWI, the fix is that decoherence of particles occurs when the particles split and begin to mingle with other particles which scatters them in the new universe causing the

> fragile quantum effect to be lost. Of course, if there is no observation of the new universes, either direct or indirect, then MWI is not a scientific theory.
>
> f> **Acceleration hedges Twin paradox:** The symmetry between views of the stationary twin and the moving twin means that each twin appears to be moving in the other's perspective. This symmetry creates the paradox that each twin is younger than the other, that is, Twin A is younger than Twin B from A's perspective and Twin B is younger than Twin A in B's perspective. The hedge to this is to say that the symmetry is broken because the moving twin (who moves away from the stationary twin and then comes back so they can compare times) must accelerate and decelerate and that somehow cancels any time dilation effects that may have accrued while the moving twin was moving at a constant (inertial) speed. Acceleration is supposed to be like gravity, both of which play with time in relativity, so the moving twin would experience time-effects due to acceleration which the stationary twin wouldn't. But, what about the period of time when the moving twin was not accelerating. And what if the moving twin just moves past the stationary twin at a constant rate of speed and they compare clocks, but the moving twin never returns and therefore does not accelerate? The paradox remains and requires another hedge.
>
> g> **Labeling rationality and logic as "common sense" rescues many mystical theories from critical analysis.** A favorite hedge used in physics is to label any reasonable critique of a theory as "commons sense." When one questions such absurdities as particles being here and there at the same time, a particle taking every possible path in going through the double slits, a cat being dead and alive at the same time, the mantra one hears is that this theory cannot be understood with commons sense. Without defining what common sense is, physicists often dismiss reasonable criticism of their theories. What they are really saying is that the theory transcends logic and rationality and is therefore mystical – while still claiming the theory is not mystical.
>
> h> **All quantum interpretations are equally valid even though they are contradictory:** Even though various interpretations of the quantum world such as the Copenhagen Interpretation and the Many Worlds Interpretation contradict each other, physicists say that the mathematics of each predict quantum phenomena equally well and are equally valid. If contradictory theories are equally predictive of phenomena, then the math must have been finagled to force the same outcome. To say these contradictory ideas are equally valid is like saying the geocentric theory is as valid as the heliocentric theory because geocentrism makes some of the same predictions as heliocentrism.

5) Qualitative analysis should precede quantitative or mathematical analysis.

In the beginning was the Word...This quote comes from a best-selling book that most of us are familiar with. I am using this quote in a different sense to say that the beginning of scientific theory is the word (i.e., language). Language must precede mathematical calculation or experiment - qualities must precede quantities. Before one can know how many there is of something, one must know clearly what kind of thing one is counting and to what conceptual domain it belongs.

Einstein is quoted as saying that mathematics is merely the bookkeeping, it's the picture that counts. I would not relegate mathematics to such a lowly estate in theory formulation; however, in stating a theory or concept, one must know the qualities (genres or categories) before assigning quantity. For example, in chemistry, qualitative analysis precedes quantitative analysis. We must know what substance is present before quantifying it. To start with the math, often leads to confusion and an inability to see the conceptual forest for the mathematical trees. As an example, a physics professor once told me that a body in constant motion does not exert force because Newton defined force as F=MA (Mass x Acceleration), so that only accelerating bodies exert force. My argument was that we must start with a qualitative or verbal definition of force. Force is something than can change the acceleration or direction of an object. Therefore, a stationary object (stationary in a given frame of reference) can change the acceleration of something that collides with it by virtue of the electromagnetic force that holds its atoms or molecules together and resists forces impinging upon it. Similarly, a body with constant velocity (inertial motion or momentum) can change the acceleration or direction of something it collides with. So given the general verbal definition of force, there is static force, momentous or inertial force and accelerated force, so that F=MA is only one expression of force even though accelerated force may be the greatest. It seems to me that the mathematical formula for force had restricted the professor's concept of force too narrowly. Again, this is an example of language (in this case mathematical language) shaping thought with a rigid semantic domain. By qualitative analysis, I mean the premises upon which a theory is based, and these include assumptions, inferences from those assumptions and concepts derived from assumptions.

Therefore, if the language and concept of a theory are not correct, no amount of mathematics can make it right as is the case with string theory.

6) Many theories of modern physics are overly mystified and can be explained in more rational, classical terms.

An excellent example of this comes from Petr Beckmann's book, *Einstein Plus Two (1987)*. Beckmann indicates that the anomalies in the precession of the orbit of Mercury were figured out by a high school teacher by the name of Gerber seventeen years before Einstein figured it out using General Relativity and tensor calculus. While Einstein invoked abstractions such as spacetime curvature and complex mathematics to explain the phenomenon, Gerber used classical concepts and simpler math to arrive at the same conclusion. While some relativists dispute this claim by Beckmann, this book you are reading will reveal many cases where classical, rational concepts can be used to explain such phenomena as the Sagnac Effect and the Michelson-Gale experiment.

7) The technical language of physics can be translated into plain, everyday language without sacrifice of accuracy as long as you define your terms.

Michio Kaku is excellent at breaking down technical language and mathematics into everyday English and making it understandable. Sometimes technical language is used sociologically to preserve the elitism of the scientific community – to obfuscate rather than communicate. This principle may seem paradoxical coming from a person who is writing about using precise language in theory building. However, many times, physics terms are less accurate than regular language in my view. The definition of space and time in relativity is a good example, and terms such as massless particles, negative energy, borrowing energy from the vacuum, etc. seem self-contradictory. Negative charge on a particle is different from the algebraic concept of negative numbers below zero. We can speak of negative temperature on the Celsius scale based on the freezing point of water, but there is no negative temperature below absolute zero. Similarly, when you reach zero energy, there is no negative energy – because "negative energy" is an algebraic, metaphysical concept superimposed on the physical world.

8) When the logical extension of a theory leads to a *reductio ad absurdum*, the theory is probably wrong.

There are a number of theories in Modern Physics, which, when taken to their logical conclusions yield absurd results. Schrodinger showed that when the Copenhagen interpretation of quantum mechanics is extended into the macro world, a cat is both dead and alive until the experimenter makes an observation. Einstein's theory of Special Relativity, when extended, falls into the trap of the twins paradox in which each twin is younger than the other depending on which point of view one assumes. The Many Worlds Theory leads to the conclusion that a new universe is born for every probability in a quantum superposition. John Wheeler, who initially encouraged his students to develop the exotic Many Worlds Theory, later said the theory carried too much metaphysical baggage to be credible. That same John Wheeler, however, believed in the Participatory Anthropic Principle in which human observation going backward in time created the universe. When a theory reduces to absurdity or leads to a paradox (contradiction is a better word), the escape hatch is always that the theory is counterintuitive and cannot be comprehended with common sense. To take that convenient escape hatch is to ignore the extensive use of thought experiments in modern physics in which paradoxes are used to disprove a competing hypothesis.

9) A thought experiment is not the same as an actual experiment.

Einstein was famous for his thought experiments or *gedanken* as they are called in German. His *modus operandi* was to start with a postulate or assumption such as "the speed of light is the speed limit in the universe – nothing can go faster than light including light itself." The next step involved setting up an apparent contradiction (paradox) to this postulate such as placing a light clock on a spaceship traveling near the speed of light. It would appear that the speed of the ship would be added to the speed of light since the ship is carrying the light source. However, since that would violate the original assumption that the speed of light is the limit, the space ship's speed cannot be added to light speed. Therefore, the idea of adding to the speed of light is reduced to absurdity based on the original assumption. However, in the rules of logic, such reasoning would be called circular because the assumption to be proved is used as proof of the assumption. So, the thought experiment essentially says that the speed of light is the speed limit and since it is assumed that the ship cannot add to the speed of light, it proves that the speed of light is the limit. Therefore, time has to slow down and space has to shrink in order for the speed of light to be the limit. The other problem with thought experiments is that they tend to lead to deductive bias when a real experiment is performed. Deductive bias leads to the temptation of filtering data so that data agreeing with the thought experiment are selected and data disagreeing with the concept are ignored. Furthermore, if mathematical calculations are performed on the thought experiment before data are collected, there is an additional bias if the mathematics seems to work. Thought experiments are useful in generating hypotheses but they should never be regarded as equivalent to an actual experiment because the pitfalls are many. Jan Hilgevoord, professor of the History of Mathematics and Science at the University of Utrecht, wrote: "I think that the figure of Einstein and the success he achieved have…seduced theoretical physicists into believing that Nature has a fixed deep structure…and that the way to discover it is to do as Einstein did: sit down at your desk and just think deeply (1995)." However, deep contemplation is pure philosophy and the beginning of science; it is not the end.

10) Zero Problem: Physicists fail to distinguish between zero balance and zero nothingness.

Here's what Michio Kaku (2015) says about the universe being created from nothing (zero) because, for example, the total negative charges are equal to the total positive charges in the cosmos, summing up to zero.

* Total matter in universe which is positive minus total energy which is negative (because gravity is negative) = 0 (i.e., heat energy repels and gravity attracts)
* Total positive charge minus total negative charge = 0
* Positive spin of galaxies minus negative spin of galaxies = 0
** Therefore the universe is zero because total matter, total charge and total spin equals zero, so - the universe could have been created from **nothing**. (Youtube Video)*

Lawrence Krauss (2012) echoes this idea that zero balance equals zero existence crediting Alan Guth as the author of this universe from a "free lunch" idea (p. 98-100). The "free lunch" idea is an obvious misunderstanding of the concept of zero. When things are in balance or equilibrium, it does not mean that there is nothing present – it simply means that opposing forces are equal. If one has an equal number of protons and electrons, it does not mean protons and electrons are not present – it means that their charges balance to net zero. Likewise, if there is a rope lying on the ground, there may be zero force being applied to each end. However, if two tug-of-war teams pick up the rope and start a game in which the force is equal on both ends of the rope (a standoff), there is net zero force, but not zero (nothing) force. Which zero force (nothing vs. net zero force) would be more likely to break the rope? The balanced tension on the rope can be measured with a scale. We will see this error in logic regarding zero manifested in much of physics writing. The following clearly differentiates zero balance from zero nothingness.

1>100 electrons + 100 protons = 0 net charge
2>0 electrons + 0 protons = 0 charge
3>100 neutrons + 100 neutrinos = 0 charge

Only in case #2 does zero charge equal zero matter (nothing). It is amazing that brilliant physicists would make such elementary errors in logic. It should also be noted that getting something from nothing is a flagrant violation of the law of conservation of mass and energy. It is amazing that brilliant physicists would make such elementary errors in logic. I think these errors come from the distinct bias against common sense (which is actually rationalism) and from the preference for the exotic and counterintuitive since the time of Einstein.

11). Confusion of zero and infinity

Gribbin (2009) in explaining the Standard Model of quantum physics, says that since particles are considered to be points with zero dimensions, then the energy between them, determined by the inverse square law, is $1/0^2$ or $1/0$ which equals infinity. The idea that one divided by nothing equals infinity crosses the border into the absurd. How could one, the smallest whole number, divided by nothing (0) become infinite? The logic or illogic goes like this. One divided by zero is usually considered to be zero, but division should be reversible by multiplication, so:

~ If $4/2 = 2$, then reversing division with multiplication $2 \times 2 = 4$ thus arriving at the original number or dividend which is 4.
~Likewise, if $1/0 = 0$, then 0×0 should equal 1, but of course it doesn't – it equals 0. Thus, the original number (dividend) could have been one, two or any number, not just the number 1. Since the original number could have been any number without end, then the answer is infinity. Of course, arriving at infinity from zero is a purely metaphysical exercise and is not related to the physical world which is the proper study of physics. Moreover, dividing 1 or any number by zero is meaningless – zero cannot divide anything.
~This conundrum can be resolved by the application of logic to abstract numbers. Logically, a number of concrete objects cannot be divided by zero – only abstract numbers can be divided by 0 in an imaginary way. To have a real division of some real objects, the divisor must be at least 2 because zero or one cannot split or divide anything.
~For example if there are 5 apples, they cannot be divided by nothing (0) - they remain 5 apples since nothing (0) cannot divide anything – it is not a real divisor. So, in that sense, contrary to our Elementary Education, 5/0 or five divided by nothing is still five - and any number divided by nothing remains that number. But, we were taught that $5/1=5$, and that is also true since 1 as a divisor does not divide or split anything either. In the same sense that we were taught that a heading of an outline must have at least two divisions, so in arithmetic, there is no division without at least a two-way split. Hence 5 apples can be divided between 2 people and each person will have 2 ½ apples ($5/2=2.5$).
~If divisor and dividend are reversed so that 0 is divided by some number, then a rational answer can be obtained. Of course, 0 real objects cannot be divided by any number. However, in baseball, if a batter comes to bat 5 times and gets no hits, his batting average is zero because $0/5$ is 0. But the average is an abstraction, unlike apples which are physically real. It is hard to imagine how nothing (0) could be divided by 5 apples or any number of anything real. To demonstrate that average is an abstraction, let's assume that the batter came to bat 5 times and got 1 hit (1/5). Now his batting average is .200, but no player can get .2 or $1/5^{th}$ of a hit. He either gets one hit or no hit – it's either 0 or 1. Again, this demonstrates that the average is an abstract, metaphysical number – albeit a useful one at times.
~ If zero is the dividend in an abstraction, then reverse multiplication is possible. To wit, $0/5=0$. Then reversing with multiplication, $0 \times 5=0$. Hence, we get back to the original dividend. However, when zero is the divisor (dividing by nothing), the operation is not reversible by multiplication because it is meaningless to divide any number by nothing (0).
~The only scenario in physics that I can conceive of where something can be divided by zero is in the entanglement theory of quantum mechanics. In this theory, entangled particles can communicate with each other over any distance instantaneously, that is, in zero time. If this theory were true, then speed would be infinite. To wit, if speed = distance/time, then speed = any distance/0 time = infinite speed. Even Einstein rejected this notion that any communication (which has to be carried by some energy) can be transmitted faster than light or electro-magnetic waves. This outlandish theory will be dealt with in a later work

> Nutshell: The zero to infinity conundrum can be resolved logically by understanding that dividing any number requires a divisor of at least 2. Zero and one cannot divide anything and zero cannot be divided by any number. The idea of dividing any number by nothing (0) or 1 is an absurdity. One (1) can be divided, but one cannot divide anything.
>
> Multiplication can be reversed, but division cannot. For example,
> 3x4=4x3 but
> ¾ not equal to 4/3,
> so 1/0 not same as 0/1.
>
> *The crux of the problem is treating zero as a number.* A number is a finite quantity of something. Thus zero, although it is an important concept and place holder in math, is not a number – it is a negation of number since it involves no quantity. It is the absence of a number as in 0 vs. 1. Likewise, there is no "infinite *number*" – that's like saying a "numberless number". The moment you add the word "number" after infinity, you have made infinity finite which is an oxymoron.

12). **Some infinities are greater than other infinities.**

This statement borders on the absurd, but some mathematicians argue that there is a difference in the size of infinities. If one infinity is greater than the other, then it would seem the smaller infinity is finite rather than infinite since it is less than the other infinity. To begin with, to use math to calculate the difference in quantities (that one thing is greater than another), there has to be two finite number of things. Here is one argument for the difference in the size of infinities.

There are infinite numbers between 0 and 1. There's .1 and .12 and .112 and an infinite collection of others. Of course, there is a bigger infinite set of numbers between 0 and 2, or between 0 and a million. Some infinities are bigger than other infinities.

However, if one can keep on adding numbers between zero and one infinitely, then the number of numbers cannot be greater no matter how large the number is that is placed above zero. In every case, regardless of number size, it boils down to 0 to infinity, because if a counter starts from zero and moves toward 1, the counter never gets to 1 to proceed to a higher number such as 2 because if he gets to 1, he has left out numbers between 0 and 1. Therefore, there is the same infinity between 0 and 1 as there is between 0 and a million or zero and infinity. Infinity cannot put an upper limit to a series of numbers – if that number is reached, then that number is not infinity. The count is always 0 to infinity no matter what upper limit we put on infinity.

One might argue that .999……ad infinitum is greater than .111…..ad infinitum, because nine is larger than 1, but again the number of digits that follow 9 would not be greater than or less than the number of digits that follow 1.

Or, one might argue that an infinity of elephants is greater than an infinity of mice because elephants are larger than mice. Here, one would be juxtaposing a finite quantity (the mass of an elephant vs. the mass of a mouse), but the numbers of elephants would neither be greater nor less than the number of mice. All we can say is that infinity always equals infinity in terms of an infinity of numbers. Of course, when one says a number, then s/he is talking about something finite – there is no *infinite number* because a number represents an end. To say that one infinity is greater than another is to say that there is something greater than infinity and that the smaller infinity wasn't infinite after all - which is a contradiction in terms.

Another argument is that a line that is infinite in one direction is greater than a line that is infinite in both directions, but the line that has a beginning is not truly infinite since it has an end point. Essentially this question is like arguing about how many angels can dance on the head of a pin. If the angels are points with zero dimensions (as subatomic particles are supposed to be according to the Standard Model), then an infinity of angels can dance on the head of a pin. The problem is that a point with no dimensions is an imaginary mathematical construct; it is not a physical reality because even subatomic particles must have dimensions if collectively they make up macro matter with three dimensions.

13) Some versions of Quantum Mechanics hold that there is no objective reality:

The Copenhagen interpretation in particular holds that there is no mind-independent, objective reality and that you cannot separate the subjective from the objective. The logic or illogic of this line of thinking is that you cannot separate the observer from the observed. If the observer alters the phenomenon observed by measurement, that alteration too is a part of nature because everything in the universe is connected including the observer – the observer does not stand apart from nature. Thus, there is no distinction made between participant observation (active observation) and detached observation (passive observation). An example of the Copenhagen line of thinking is: You cannot know the position and speed of a particle at the same time because when you do something to the particle to detect its location, you change its speed. Since you cannot know the position and speed at the same time, then they don't exist at the same time in nature. The problem is that classical science requires objective evidence to support a theory, and the theory of evolution is based on the idea that objective physical reality came into existence long before human consciousness. In order to get around this problem, Copenhagenists have to say that human consciousness going backward in time created everything that went before humans. Thus, by extension, everything that is created in the future must be created by human consciousness even if the human species become extinct. How can anyone maintain this point of view without severe cognitive dissonance? The answer is always the mantra: quantum mechanics cannot be understood with common sense (rationalism is a better word).

Applying my Principles in a Debate with a Physics Student

Student: "You have no right to challenge physics theories because you do not have a degree in the field."

Me: "You may be right, but let's take an example such as String Theory. Is it true that for String Theory to work, there has to be 11 dimensions of space?"

Student: "Yes."

Me: "Is there one shred of empirical evidence for any extra dimensions, i.e., any dimensions other than the three that we experience every day plus time?"

Student: "No, I don't believe there is."

Me: "Then don't you see that the layman can question the assumptions, logic, language and inferences drawn from the data and that String Theory hangs by a very thin thread. If the assumption about extra dimensions is wrong, then the theory has to be wrong."

Student: "Yes, but the mathematics that come out of string theory is so elegant and intricately consistent that it must represent reality."

Me: "But the idea of extra dimensions is a mathematical construct, and a mathematical proof is not the same as a scientific proof. Take the game of chess. You can find some complex mathematical regularities in chess moves and even program a computer to beat the world's chess champion, but do those mathematics represent any physical system in the universe that you know of?"

This little debate with a student of physics goes beyond string theory. The fallacious-type of thinking that I revealed is symptomatic of the problem in Modern Physics. We see that metaphysics and abstract mathematical constructs have largely replaced empiricism as the key to proving theory. Several physicists have come to the same conclusion. More later on that.

So when can a philosopher or layman have a legitimate opinion in physics?

The above dialogue represents my belief that the informed layman can critique scientific theories if s/he knows the language of the scientific method and the philosophy of science. When the author of string theory posits that there are seven extra dimensions that are not empirically validate-able, then he is entering the realm of philosophy, and others can have a say on the issue based on epistemology. If extra dimensions are metaphysical assumptions made by mathematicians, they can certainly be questioned on the grounds of empiricism, the hallmark of scientific thinking. I do not have to understand the intricate calculus involved in String Theory to know that it is based on shaky assumptions about the empirical world. As stated, a scientific theory begins in language and logic – it should not begin in mathematical language. Thus, the layman can question the hypothesis, the assumptions and the inferences drawn from the data. The math can only be as good as the assumptions and the theory that underpins it. We all know that mathematical regularities can be created in artificial systems (such as chess) that have little or no resemblance to the empirical world where science resides.

CHAPTER 1: LANGUAGE, LOGIC AND THEORY

But the Lord came down to see the city and the tower the people were building. The Lord said, "If as one people speaking the same language they have begun to do this, then nothing they plan to do will be impossible for them. Come, let us go down and confuse their language so they will not understand each other." (Genesis 11:5-7 New International Version (NIV)

The above quote from the Bible is not meant as a theological statement, but it is intended as a metaphor for comparing the confusion of language which, in the Biblical story, prevented the building of the Tower of Babel (pronounce it "babble"), to the confusion of language that often prevails in the building of scientific theory. In my study of anthropology, I was most intrigued by the sub-field called linguistics. In linguistics, I learned that language is not only a medium of communication, but that language can influence the way people think and perceive the world. Thus, language is the vehicle that carries the philosophy of a culture and implicitly embeds ideas about what is real, good and beautiful.

Since the human mind has a limited digit span, every language has to break up the continuum of reality into units to form categories of thought, and those categories, which vary from language to language, influence the way people mentally construct the world around them. This concept has come to be known as the Sapir-Whorf hypothesis of linguistic relativity – yes, anthropology has relativity too, including cultural relativity. As an example, Zuni Indians divide the color spectrum into categories different from English speakers. Zunis combine yellow and orange into one category and give it one name (Lee 1996). However, Zunis are not color blind – they have the same rods and cones in their retinas as other groups. They can sense the difference between yellow and orange, but they see them as different shades of the same color. Hence, we can say that the Zuni language does not influence the Zuni speaker's *sensation* of color, but does influence the Zuni's *perception* of color. To wit, Zunis are not as sensitive to this yellow-orange difference as English speakers who are very color conscious by virtue of a language rich in color terminology. Therefore, when a test was given to some Zunis and English speakers sorting yellow and orange balls of yarn, the Zunis made significantly more errors than English speakers. Not only do noun categories such as colors influence thought, but grammatical structures also influence cognition and abstract thought. Furthermore, there are linguistic differences in the perception of time with some cultures conceptualizing time as a circle and others seeing it as a straight line.

Since we speak of a stream of consciousness and we link thought to language, one might infer that language is a continuous stream as well. However, language is *quantizable* in a metaphorical sense in that the elemental unit of sound is a "phoneme." A phoneme is a difference in sound that makes a difference in the meaning of an utterance. For instance, you place your tongue to the roof of your mouth and aspirate (blow air from behind the tongue) when you pronounce a "t", but you put your tongue in the same position but do not aspirate, when you pronounce a "d" which is also vocalized. Hence "t" is a different phoneme from "d" because it makes a difference in meaning. For example, if you say "tab" and then you say "dab", two different images are evoked in your mind if you are an English speaker. Phonemes combine to form morphemes or units of meaning (roughly equivalent to words), and morphemes combine to form larger units of meaning we learned when we studied grammar. The study of linguistics is considered to be a science (albeit a soft science), and the theory of linguistic relativity is a key in the evaluation of special and general relativity as I shall attempt to show.

Some linguists have been intrigued by the possibility of consciously creating new artificial languages that could empower more precise ways of thinking. Natural languages have evolved by trial and error and by the amalgamation of different languages and thus contain many inconsistencies and grammatical anomalies. English is a good example of an amalgamated language made up of Celtic, German, French, Nordic and Latin roots. It has a German grammatical structure with many Latin root words derived from the Roman invasion. Thus, there are many exceptions to grammatical rules in English.

An example of an "artificial" language designed to maximize human intellectual potential includes "Loglan", specifically designed by its creator James Cooke Brown to test linguistic relativity, by experimenting with whether a more precise language would make its speakers think more logically. Speakers of "Lojban", a development of Loglan, report that they feel speaking the language enhances their ability for logical thinking (Brown 1975). Kenneth E. Iverson, the originator of the APL programming language, believed that the Sapir-Whorf hypothesis applied to computer languages (without actually mentioning the hypothesis by name). His Turing award lecture, "Notation as a tool of thought", was devoted to this theme, arguing that more powerful notations aided thinking about computer algorithms (Iverson 1980). Of course, the idea of designing a near-perfect language is frustrated by the digit span of the human mind. Psychologists tell us that we can hold only about 7 units of information in the mind at once. However, we can increase our digit span by "chunking" together several pieces of information into one category and that category can become one unit of information. For example, to attempt to remember the number 14921776911 after looking at it once would be impossible for most people because it exceeds their digit span. However, it we chunk the numbers into 1492, 1776 and 9-11, most Americans could remember these three chunks. Accordingly, because of the limit of our RAM, or short-term memory, language must break reality into gulp-size bites of about seven items held in our heads at once. Then these seven items or chunks must be sequenced with words in a grammatically-prescribed order. The speed of thought is limited by the electro-chemical process of neuronal transmission which is much closer to the speed of sound than the speed of light. Until our brains can grow larger or more efficient, we are neurologically limited in terms of language and thought – although computers can aid us in processing data much faster than the human brain can process it.

In reading physics and talking with physicists, I have come to realize that physics, like all other academic disciplines, has developed its own specialized culture and language with dialects which contain assumptions and concepts about what is real in the world. Culture is something that naturally develops when individuals interact in a group. Although there exists diversity in a group, a core of beliefs and norms will emerge on which most of the members agree. As the group grows larger, subcultures begin to form with some variation from the mother culture. Language is the vehicle that carries the cultural ideas and theories about reality and, once formed, places constraints on the way the group sees the world. Classical physics is one dialect of physics language whose way of seeing the world is more intuitive and based on our common sense about the way the slow, macro-world works; relativity is yet another dialect that deals with another aspect of reality (the very fast and the amount of gravity present); and quantum physics is still another dialect that describes the world of the very small in ways that are counter-intuitive and thus do not accord with the way our senses and perceptions have been designed to see the world. The problem arises when these dialects of physics do not agree with each other. Paradoxes, if not outright contradictions, are created when attempting to translate classical physics into relativity and relativity into quantum theory.

Can Language actually make a difference in the accuracy of a theory?

The whole theory of relativity (Special and General) hinges on the definition of space and time. If space and time are not physical entities that can interact with mass and energy, as Einstein assumed, then it would seem that the whole theory collapses. However, it could be argued that if spacetime is a metaphor for the way matter and energy change as they interact in space over time, then it could yield accurate predictions even as Ptolemy's geocentric theory, buttressed by epicycles, yielded rather accurate predictions. Here is what physicist Daniel Styer has to say about the relationship of language to theory.

The English language was invented by people who were not familiar with relativity, so it often works against our understanding. There's a word for "before" and a word for "after," but there's no word for "maybe before or maybe after or maybe at the same time, depending on reference frame." Our commonsense notions of space and time have been built into our language, but our commonsense notions of space and time are wrong. We have to be on our guard because our language encourages us to say (and to think!) false things (2011: Kindle Locations 764-769).

Of course, Styer is saying that the English language is not an adequate vehicle for conveying the concepts of relativity. I would say that the language of relativity leads us astray because the theory does not attempt to define time and space, and its incorrect use of these terms leads to outlandish and contradictory conclusions. To correct for these so-called paradoxes (contradictions is a better word), relativists have to create more faulty language in attempting to correct the uncorrectable. Here's what physicist-mathematician Peter Woit (2006) had to say about the loose use of language by physicists based on his dual identity as a mathematician and physicist:

To mathematicians, what is at issue here is how strongly to defend what they consider their central virtue, that of rigorously precise thought...Physicists have traditionally never had the slightest interest in this virtue, feeling they had no need for it. This attitude was justified in the past when there were experimental data to keep them honest...(p. 260).

Could a more precise language bridge some of the gaps among the classical world, relativity and quantum theory? I think so. Then there is the language of mathematics (quantitative language) which has come to dominate physics. Furthermore, Eastern philosophy has crept into physics as Fritjof Capra has noted, so that paradoxes or contradictions can coexist in the same theory. How can a particle be said to be massless whether we are talking about a particle of matter or packet of energy? Logically, a particle without mass (even energy has mass) is not a particle. Particles with zero mass could never add up to an object with mass.

> The arbitrariness of language means that the vocabulary that represents the external world is arbitrary. For example, the plant we call a "tree" is represented by a different sound in different languages. In Spanish, the word for tree is "arbol". Furthermore, the way we categorize objects in the world is somewhat arbitrary. This arbitrariness is also true of mathematical language in the sense that mathematical language is, like qualitative language, an epiphenomenon that we superimpose on the world in attempting to describe and understand it. Math is not somehow innate in nature.

Some Imprecise Terms Used in Physics Somewhat Clarified

Yogi Berra, famous for his many miss-speaks, tried to disavow some of the things attributed to him by saying that "I didn't say everything I said." Perhaps, physicists might say: "We didn't literally mean everything we said." In linguistics, **semantics** is the study of how we attach meaning to words. The ideal of scientific semantics, unlike poetry and literature, is that each word should have one and only one literal, precise meaning. In other words, "nothing" should literally mean no-thing, vacuum, zero, *nada*. However, in much of physics parlance, "nothing" doesn't mean exactly nothing or zero, and negative numbers do not always mean less than zero or nothing. Here are some terms that are used in physics, but their meaning in physics isn't exactly what they mean in Standard English

1) **Vacuum** means there is no matter including gas in a given volume, but radiation in the form of photons vibrating at various frequencies is present in the vacuum. Thus light, electro-magnetism and neutrinos are widespread in the universe – even in areas called "the vacuum". Of course, the fallacy of this concept is that matter and energy are treated as separate entities when the equivalence and interchangeability of matter and energy is a well-accepted dictum in physics.
2) **Zero point energy** – zero energy does not mean no energy in physics. Zero-point energy is the minimum known energy or radiation in a given volume of space. Perhaps the void in intergalactic space would contain low levels of radiation and that would be the lowest level of energy in the universe, but it would not be zero energy in the literal sense of zero as nothing.
3) **Virtual photons** popping up out of the vacuum and disappearing – I would say the photons were already there from radiation spewed into space by trillions of stars. There is a surge of the radiation at certain times and places – much like a surge of atmosphere that we call wind. The air was there before it started moving, and the photons were there before they surged and manifested themselves – they did not pop up out of nothing as physicists often say. But then, again, nothing is not really nothing in physics semantics.
4) **Massless particles**: photons are said to be massless and the Standard Model is divided between particles with mass and particles without mass. Einstein and Bohr certainly thought that photons had mass – probably because it is known they have energy and energy is equivalent to mass according to $E=MC^2$. When Einstein created a thought experiment to challenge the Heisenberg energy-time complementarity, he imagined a clock-scale with a light-emitter – all suspended from the ceiling. When the clock emitted a photon, the scale would register a loss of mass which would be recorded on the clock instantly. Bohr became very upset by this thought experiment, but he never questioned that the photon has mass. He finally rebutted the argument, by evoking General Relativity, which would predict that the exact time could not be known to all observers since gravity affects time flow, so that there would be no simultaneity between the clock and

observers in different gravity fields.

4) **Spacetime curvature** is defined as a physical entity that can affect and be affected by matter-energy. Spacetime is an abstraction, not a physical entity.

5) **Extra dimensions** – There is no evidence for extra dimensions – they are mathematical constructs which Einstein initially rejected.

6) **Anti-gravity or negative gravity**: Negative gravity in come contexts does not mean below zero gravity. It simply means that negative gravity would be repulsive rather than attractive and would involve negative inertia. However, "negative gravity" is not only used to mean repulsive gravity but is used to mean less than no gravity. For example, as we shall see later, some physicists claim that since negative gravity is equal to positive gravity, the universe could have come from *nothing*. Of course, the net balance of negative and positive gravity is not the same as no gravity where there is no mass to create gravity.

Here is William Mitchell's critique of the use of the terms "nothing" and "negative" in Modern Physics.

Although any curvature of empty space is illogical, the negative curvature of space is ridiculous. That would require the presence of negative matter to produce negative gravity, and negative gravity to produce negative curvature, all of which are pipe dreams. Opposing forces may be labelled positive and negative for mathematical purposes, but all are actually positive (2002: p. 234).

Language and Mysticism

Science does not need mysticism and mysticism does not need science, but man needs both. – (epilogue to the Tao of Physics by Fritjof Capra)

Does modern physics (relativity and quantum mechanics) require a different kind of logic based on metaphysics or perhaps mysticism? Some physicists say "yes"; others vehemently deny that modern physics has deviated from the classical scientific method. It seems to me, to say that the theories of modern physics are counter-intuitive and beyond common sense rationalism is to admit that modern physics requires a different logic – perhaps a fuzzy logic or mystical logic – if such qualifies as logic. Perhaps I should say that Modern Physics, like mysticism, requires a different thought process which is *paralogical* or perhaps non-logical. Fritjof Capra (1999) and David Bohm (1980) had much to say about the relationship of language and mysticism to theory. Both Bohm and Capra were admittedly given to mysticism, and Capra wrote a famous book (*The Tao of Physics*) detailing how modern physics follows some of the same thought processes as Eastern mysticism. Essentially, both these theorists said that reality is an indivisible whole and when one tries to break the continuum of reality into categories of language, she distorts the true nature of the universe. Thus, Capra indicated that modern physics transcends language and must be understood through mysticism - and that Eastern mysticism is the best vehicle for conveying physics theories. Bohm, who was mentored in Eastern philosophy by Jiddu Krishnamurti, echoed this thought by saying that the truth of the universe is beyond language because it is a "seamless whole" with a deep structure that he called the "implicate order." Krishnamurti's spiritualism "helped Bohm put his idea of the oneness of the universe into a philosophical context (Clegg 2014: p. 89)." Additionally, Bohm saw linguist Noam Chomsky's innate "deep structure of grammar" as a manifestation of the implicate order. This is somewhat ironic since Bohm believed that ultimate unitary truth is beyond language categories.

However, it is interesting that both Capra and Bohm wrote extensively using language categories to describe this universal oneness. What else besides language, including mathematical language, do we have to describe and analyze the universe? Contemplating your navel and its oneness with the universe may yield some mystical insights, but it won't get you very far in understanding why your navel and you orbit the sun and why the sun orbits the galaxy. As Jim Carrey put it in a silly but humorous movie in which he is contemplating leaving a Buddhist monastery, "But I have yet to achieve *omnipresent, super-galactic oneness*" (Quote from *Nature Calls 1995*). To which the monk, eager to get rid of the Pet Detective, answered: "Oh you are *one* – you may go." Ironically, Jim Carrey now has a series of YouTube videos on how to achieve happiness through finding oneness with the universe. Silly movies aside, Capra says that modern physics, like Zen Buddhism, presents the initiate with riddles or *koans* designed to break him out of the limits of rational thought to enable him to embrace the mystical oneness. In Zen Buddhism, a favorite koan is "What is the sound of one hand clapping?" Of course, there is no rational answer to this riddle. The riddle sets up a contradiction that cannot be resolved logically: one hand can't clap and therefore makes no sound at all (logical answer) vs. the assumption in the question that one hand can clap and make a sound (non-logical answer). Since the koan cannot be logically resolved, the disciple is forced to open his mind to mystical, non-rational, transcendental consciousness where there are no contradictions or dualities because everything is one. Capra says that physics also uses koans that challenge our rational, linguistic understanding of the world. Let me count the koans: a particle must spin twice to return to its original position, a particle can be here and there at the same time, massless particles, negative energy, negative mass, borrowing energy from the vacuum, a particle takes all possible paths in going through the double slits, a cat can be dead and alive or neither dead nor alive until an experimenter makes it one way or the other by observation.

The inimitable Yogi Berra may have something to contribute to the connection between Eastern mysticism and modern physics. Yogi was quoted as saying: "If you come to a fork in the road, take it." Perhaps that is the reason he was called Yogi, because his "Yogiisms" resembled the mystical, contradictory sayings of real Yogis of India. Maybe the "fork-in-the-road" saying could be applied to quantum quandaries where, in one interpretation, the particle takes every possible path in going through the slits. So, the particle takes every fork in the road to which it comes *a la* Yogi Berra. However, if Yogi were still alive, he might deny some of the sayings attributed to him as he did on one occasion when he said: "I really didn't say everything I said." Hereinafter, I shall refer to this kind of logic or illogic as "Yogi logic". Yogi's misspeaks or mystical-speaks aside, mysticism certainly has its place in literature and art in which we are invited to a different perception of the world as in Dali's surrealism. We also witness mystical metaphors used in poetry and song where they are often called "conceits." Thus Billy Joel can sing "This night will last forever" knowing rationally that no night literally lasts forever even though a romantic, enchanted evening may linger in memory for many years to come.

So mysticism has its place in the humanities including religion, art, music and literature, but not in science, and that is the reason that our universities divide the curriculum into the humanities and sciences - sometimes called the arts and sciences. Mystical and religious concepts seem to be beyond the scientific domain and should reside in the "humanities." However, the scientific method can, in some cases, be used to conduct research into the paranormal. Such things as near-death experiences, the anthropic principle, or the role of the observer in affecting material reality, and other questions, which are more akin to spiritual matters, can be studied scientifically if there is a claim for physical evidence of these phenomena. However, such research should be done with the same rigor as is used in the hard sciences. It is the position of this author that the discipline of philosophy is the mother of both humanities and sciences and is the bridge between the two.

Other physicists have recognized the mysticism inherent in quantum theory. Neils Bohr famously said: "If quantum mechanics hasn't profoundly shocked you, you haven't understood it yet" (Brainy quotes). However, I agree more with what Richard Feynman said: "I think I can safely say that nobody understands quantum mechanics" (1965: *The Character of Physical Law*). It would seem that when the human mind is confronted with a mystery, the mind turns to mysticism. It is interesting that the word "mist" is related to mysticism. In the "mist," it is difficult to see the world clearly because of the water vapor in the air.

Robert Ornstein, psychologist of consciousness, has suggested that the right hemisphere of the brain is the center of mystical, non-linear, non-rational and non-linguistic thought. In doing so, Ornstein echoes many of the ideas of Capra and Bohm about the mystical, unitary nature of reality and draws much of his inspiration from Sufism (Islamic mystical traditions). In Ornstein's words:

The division of the mind is profound, and it begins earlier than we had thought, not in early human society, not in our remote humanoid ancestors, not in monkeys, but before primates (1997: p. 4).

The following is my amplification of Ornstein's voluminous work on consciousness as it relates to left brain and right brain thinking. Ornstein is particularly interested in the intuitive functions of the right brain which he feels has been neglected in Western thought because of the silent hemisphere's tendency toward mysticism. The following distinctions are admittedly oversimplifications, but they show tendencies in the differential functioning of the two hemispheres.

Left brain is to right brain as

Western is to Eastern philosophy*
Language is to non-linguistic thought
Logic is to intuition
Linear is to non-linear/simultaneous
Rationalism is to mysticism
Time is to space
Part is to whole
Analysis is to holism
Individuality is to oneness
Classical physics is to Modern Physics

*Of course, we know that the global distinction of Eastern and Western is overly simplified. There are mystical traditions in the West (including Christianity), and there is scientific, rational-empiricism in the East, especially in the modern world. In Western Civilization, which is closely associated with Christendom, the trinity is a mystical concept in which three Gods are somehow united as One God, and Jesus is seen as completely human and completely God at the same time. Nevertheless, perhaps there is historically more mystical thinking in the Eastern hemisphere of the globe and in the right hemisphere of the brain. Ironically, from a Eurocentric point of view facing north on the globe, the West is to the left and the East is to the right.

After writing this analysis of Modern Physics in terms of right brain and left brain thinking, I encountered an author who agrees with my thesis. Gary Zukav, author of The *Dancing Wu Li Masters* says that:

Roughly speaking, the left hemisphere is "rational" and the right hemisphere is "irrational." The Copenhagen Interpretation (of quantum physics) was, in effect, a recognition of the limitations of left hemispheric thought…It was also a re-cognition of those psychic aspects which long had been ignored in a rationalistic society (p. 42).

Thus Zukav is saying that quantum physics is right-hemispheric and irrational and therefore beyond understanding by the rational mind (the left hemisphere which is the seat of language). Like Fritjof Capra, Zukav says that Eastern mysticism is a better vehicle for understanding quantum physics than Classical Western science, and he uses the Dancing Wu Li Masters as a right hemispheric, intuitive way of understanding Modern Physics. Dancing Wu Li Masters, according to Zukav, contend that the universe is *organic*, that is, it is a living organism that is integrated and unified and cannot be separated into parts. Thus, an understanding of the world is beyond language with its limiting categories and must be danced out rather than described in words. In the dance, the dancer and the dance are one. As in the Copenhagen interpretation, the observer cannot be separated from the observed - just as the dance cannot be separated from the dancer. I would critique this kind of thinking by saying that the dancer and the dance are one because it is the dance that links them even though they are different entities. The dancer could also sing, paint, play an instrument, do science, etc., and the same dance could be performed by another dancer. The fallacy is to say that unity means sameness rather than linkage. However, in my survey of physicists, most denied that Modern Physics requires a different kind of logic or thought process to understand it and thus deny any connection to Eastern mysticism.

Is Mystical Unity Based in Diversity or is Unity based on Sameness

What Zukav, Capra and Bohm fail to clarify is the distinction between "difference" and "separateness". Is reality *one* in the sense that everything is the *same*, or is it *one* in the sense that, although there are differences between the parts, the different parts are connected to each other in a larger *whole*? It is interesting to speculate as to exactly what *oneness* means in the mystical traditions. There are two possibilities: 1) Oneness means everything is exactly the same and the perception of difference is an illusion. As Zukav puts it, reality is an undifferentiated whole. Although the surface manifestations of reality may appear different, beneath the surface there is one undifferentiated reality – and that is the true reality. 2) There are differences in phenomena, but at a deeper level, these separate parts are connected into a larger whole. Here, one might invoke the concept of the American melting pot which is *e pluribus unum* (unity in diversity). In quantum physics, distinctions are indeed made among the particles that make up the atom. Protons, neutrons, quarks, electrons and photons have their separate identities and can exist independently of each other, yet they can combine to make up a whole which is the atom. Also, cosmologists argue that at the time of the Big Bang, there was undifferentiated whole, but as the universe cooled, distinct, separate particles condensed out of the foam of energy. So, all particles are connected genealogically to the original stem particle that condensed out of a cloud of energy. Similarly, the cells of the body are differentiated as skin cells, nerve cells, muscle cells, etc., but they are all connected and integrated in the body and serve specialized functions in the survival of the organism. Certainly, we would not want to say that all cells are alike even though they may begin as undifferentiated parent stem cells.

At the macro level, let's take the differences between men and women as they relate to marriage (and this is not a political polemic against other forms of marriage besides the male and female kind). Men and women are obviously different in their plumbing and wiring, and recent research has shown that there are average differences in their psychology as well. Yet, on a deeper level, men and women are connected by a common humanity – they are members of the species, *Homo sapiens*, and they get part of the DNA from a male (father) and part from a female (mother). When married, husband and wife are said to become one. That does not mean that they have merged as one Siamese-twin entity or that they become identical to each other. The idea is that they are connected in a psychological or spiritual sense, and they should ideally pursue unity so that their differences will complement each other. A divorce rate of over 50% in the US indicates that the ideal of unity is just that – an ideal. Married couples discover that there are differences, and these differences are not just individual in nature – many are rooted in gender. In the Eastern concept of yin and yang, differences do exist, but differences complement each other in forming a larger unity. Hence there is a difference between a) a paradox where differences may merge into a larger synthesis and b) a contradiction which involves true opposites that cannot be resolved into oneness. Males and females are different but not absolute opposites since they have much in common as humans, but a cat cannot be dead and alive at the same time. A dead-alive cat is an opposite that cannot be resolved.

Perhaps the best synthesis of Eastern and Western, right hemispheric and left hemispheric thought is to be found in the adage: "the whole is greater than the sum of the parts". However, I would add to that saying: "the whole does not exist without the parts". The holistic part of the saying is more of an Eastern (right hemispheric) way of looking at the world, and the fact that the whole does not exist without the parts is more of a *sine non qua* or "not without which" Western (left hemispheric) way of viewing the cosmos. So, the two concepts combine to form a synthesis of East and West. Maybe we could trace this structure of thought to the structure of the brain itself. The left hemisphere (seat of language and analytic thought) is merged with the right hemisphere (seat of holistic thought) by the corpus callosum (the bundle of nerves that connect left and right hemispheres). Perhaps the corpus callosum itself is the place where the twain of East and West meet. The left and right hemispheres do not have to be the same to form a unit, but they do have to be connected to act as a unit. After all, if they were the same, they would not need to be connected to achieve complementarity. Furthermore, the search for a unified field theory is physics is a right-hemispheric, Eastern way of thinking. However, a unified theory such as electro-magnetism does not mean that electricity and magnetism are exactly the same, but that they are separate manifestations of a common force and intimately connected to each other. Even if the *string* is the undifferentiated unit of matter, strings must take on different shapes and different vibrations to create the zoo of particles physicists observe. The stem cell is an undifferentiated cell that can transform into any kind of specialized cell, but it has to transform into a human cell, retaining the 46 chromosomes of human DNA.

Dualistic and Monistic Philosophy in Physics

Dualism seems to be rooted in the deep structure of the brain as evidenced by the widespread, if not universal, presence of dualistic thinking in cultures around the world. Likewise, the tendency to synthesize dualities and resolve the conflict between them seems to be rooted in the brain as indicated in the more or less universal plots of mythologies around the world which involve duality>conflict>resolution. Hegel said it well with his concept of how ideas evolve through the processes of thesis>antithesis>synthesis. Similarly, we see dualities in physics theory which may have as much to do with the structure of the brains of physicists as the nature of physical reality. Of course, if one takes an evolutionary point of view, then the mind must have evolved to mirror reality to some extent and to manipulate the physical environment. Consider these dualities.

Matter vs. anti-matter
Mass vs. negative mass
Energy vs. negative energy
Gravity vs. anti-gravity
mass particles vs. massless particles

Now duality can sometimes mirror nature and sometimes not. There are times when the mind can create a false duality that does not represent nature. For example, can there by a massless particle when all matter and energy have mass according to Einstein's famous equation ($E=MC^2$)? Can there be negative mass in the sense of mass that is below zero or no mass? Some physicists distinguish between matter and mass indicating that matter is measured by the number of moles of atoms in a substance as there are atoms in 12 grams of carbon-12. Mass, on the other hand, is measured by inertia or the resistance to acceleration. The faster matter is moving, the more inertia it has even though the number of moles or atoms in the substance does not change. Movement imparts to a substance virtual mass, but no more matter. Hence mass (inertia) is a property of matter – it can never exist separate from matter. Therefore, there is no mass without matter, and there is no matter without mass. Ergo, the idea of a massless particle or negative mass is a false duality. If inertia (mass) is an inherent property of matter, then it cannot be separated from matter and still have matter. Therefore, the idea of negative mass and massless particles is not synthesizable in the Hegelian sense and do not reflect nature. Also, the idea of anti-gravity which involves negative mass, negative inertia, dark vacuum energy does not follow scientific realism and is unscientific. However, if one means that there can be a repulsive force that is stronger than gravity, then that could be a scientific reality. For example, a magnet which lifts an object against the force of gravity would overcome the inertia of that object and in a sense provide anti-gravity, but it does not mean the object has lost its inertia or mass and certainly not its matter. Consider this conundrum:

matter>zero matter>negative matter
mass>0>negative mass

Obviously, negative matter is an imaginary thing and any number assigned to it is an imaginary (and therefore physically unreal). Now some physicists argue that if an object, such as a hot air balloon, is lifted to a certain height and hovers, the air pressure equals gravity therefore gravity equals zero. Or one could argue, mathematically, that as the balloon is rising, there is negative gravity because air pressure is greater than gravity. This line of reasoning confuses mathematics with reality and is a poor use of language and logic. A zero in mathematical calculation does not mean that there is no gravity acting on a hovering balloon, it means the two forces are balanced, and if there were no gravity or less gravity, the balloon would continue to rise.

To illustrate the fallacy of this kind of thinking, I will use a previous example. Consider a rope that is being used in a tug of war. If the two competing teams are exerting equal force on the rope and neither is winning, then the net force is zero because the forces are balanced. Likewise, if the rope is lying on the ground and no one is pulling on either end, the force is also zero. Yet, who would argue that these two zeroes are the same thing. Obviously, the rope being used in a tug of war has much more tension applied to it and is much more likely to be stretched or broken than the rope on the ground. There are many cases in physics, where a zero in mathematics is mistaken for a zero quantity rather than a zero balance. Such is the case in which Alan Guth who explains how the universe is made from nothing because so many factors balance out to zero in mathematical cosmology. Here is Lawrence Krauss's paraphrase of Guth in Krauss's book *A Universe from Nothing*.

(The) ultimate "free lunch including the effects of gravity in thinking about the universe allows objects to have— amazingly—" negative" as well as "positive" energy. This facet of gravity allows for the possibility that positive energy stuff, like matter and radiation, can be complemented by negative energy configurations that just balance the energy of the created positive energy stuff. In so doing, gravity can start out with an empty universe— and end up with a filled one (Krauss. P.99).

Again the fallacy in this "universe that created itself from nothing" thinking is that a balance of positive and negative energy means zero energy, rather than zero net energy. To the contrary, it means that there are equal amounts of positive and negative energy. Having an equal number of electrons and protons does not mean you have no protons and no electrons. If you have two protons and two electrons, you have four subatomic particles whose charges are balanced to net zero charge – not zero particles. This seems like elementary logic, but I see it frequently in reading physics. Of course, what Krauss is trying to prove in his book is that you can get something from nothing as a natural process without somehow violating the law of conservation of mass and energy. Krauss is making a theological (or anti-theological) statement that an intelligent designer (deity) is not needed to get a universe from nothing. Debating theological issues is not the purpose of this work, however. Those issues are "undecidable" from a scientific point of view as Michio Kaku has said and is a matter of personal faith outside the scientific domain.

The fact that mystics seek universal oneness that passes all understanding seems to partake of the same psychological need that drives physicists to achieve the elusive dream of a unified field theory aka TOE (Theory of everything) or GUT (Grand Unification Theory). Part of that dream is to unify the dualities that physicists have found from Newton's opposite and equal forces to the wave-particle duality to the notion that for every particle there must be an opposite and equal anti-particle. Just as Chomsky (1957) suggested that there was an innate deep structure in the brain that underlies the grammar of all languages, there must be an innate structure that compels humans to seek unity in diversity and to transcend dualities. In Hinduism, there are many gods, but they are considered mere manifestations of the one Brahmin. As a result of Eastern influence, Bohr adopted the yin yang symbol as part of his family coat of arms when he was knighted in 1947 (Gules and Sable: Escutcheons of Science). Since the yin-yang symbol represents the complementarity of opposites, Bohr was expressing his belief in the unity that transcends duality and the unification of the forces of nature.

Does the search for symmetry and unification of dualities say more about the structure of nature or the structure of the human mind? Claude Lévi-Strauss (1967), French anthropologist, argued that the "savage" mind had the same structures as the "civilized" mind and that human characteristics are the same everywhere. Perhaps the human mind has evolved (or been created) so that its structure is similar to the structure of nature. The idea of innate ideas spans the gamut from Greek philosophy to modern philosophers and psychologists such as Carl Jung. Be that as it may, I believe that the human mind is structured in such a way as to represent, but also to oversimplify nature and to manipulate nature in the development of technology as a survival mechanism. Thus, to achieve a deeper understanding of nature, we must strive to transcend the simplicities of monism and dualism. Nature is never quite as simple and neat as our theories as Feynman and Capra have argued in saying that our theories are approximations of reality. Capra *(1999)* said it well:

Modern Physics has confirmed most dramatically...that all the concepts we use to describe nature are limited, that they are not features of reality...but creations of the mind; parts of the map, not of the territory (p. 161).

Perhaps there is a synthesis between dualism and monism as Hegel has suggested in what has become known as an Hegelian synthesis in which thesis vs. antithesis leads to a new synthesis which in turn becomes a new thesis to be opposed by anti-thesis and so goes the cycle of thought. Hopefully physics theory will grow through such synthesis of ideas and not be too territorial in rejecting ideas from outside the field.

Critique of Bohm and Capra

As an anthropologist, I also see many identities between the philosophy of modern physics, Eastern mysticism and the cultures of primal peoples. The Australian Aborigine concept of "dreamtime" bears a great deal of resemblance to the concept of time and various dimensions in modern physics.

It is also interesting that both Bohm and Capra were quantum physicists. The essence of quantum physics is that energy is not continuous and indivisible as once assumed but is made up of discrete packets or units called "quanta" as described by Planck's constant. How can the reality of physics be a "seamless whole" if the material universe is made up of quanta, atoms, subatomic particles and other distinct, separate units which combine to form the material world that we see? Even these units we call atoms can be split releasing even finer particles and deadly radiation.

Bohm (1980) wrote rather extensively on the relevance of language in theory formation. Much of what he said mirrors the Sapir-Whorf hypothesis that language shapes thought (p. 17). His thesis is that English, with its emphasis on nouns, gives a fragmented view of the world. The sequence of subject-verb-object separates the actor from the action, subject from object. For example, in English, we say "It is raining". What does the "it" in that sentence refer to? Why do we need a noun? Why not just say "raining" as a Hopi Indian might say? My comment is that "It is raining" is more vernacular than formal speech. To be more scientific, we might say that "evaporated water in the atmosphere has reached the condensation point." Then the pronoun "it" refers to some process that can be described more precisely. Bohm, who was raised as a Jew but rejected the faith of his heritage at a young age, goes on to say that ancient Hebrew is a verb-oriented language where all parts of speech and grammatical forms are derived from verbs. It is a dynamic language that unites actors and objects in a seamless whole. So, just as Walter Goldsmidt (1970) says that the Hopi language is more suitable to describe dynamics of physics because of its process orientation (Lecture at Wake Forest University), Bohm contends that ancient Hebrew is a better vehicle for conveying physics.

My critique of Bohm's thesis is that nouns and verbs go together like a horse and carriage. The grammatical structure of English actually unites the noun and verb, subject and predicate, actor, action and acted upon. Bohm's fallacy is that oneness is sameness (actor, action and object are the same) when a better view of oneness is that actor and object are connected by an action – although actor and object are different. Thus, they are separate but connected into a unity (*e pluribus unum*).

Logically, for there to be an action, there must be an actor, and there has to be an object of an action verb. With only verbs, there is a disembodied ghostly process without actors or objects. If I say "hitting" by itself, the first question you would ask is what or who is hitting and what or whom are they hitting. The verb alone yields only vagaries. The universe is indeed composed of hard matter (nouns) that move, exert force and give off energy (verbs). It is difficult to think of any force or energy that is not intimately associated with matter. If you want to create some light or make an atomic bomb, you must start with some matter. Hence, we could think of matter as being the nouns of the universe and energy and force as providing the verbs or action in the universe because energy (which is derived from matter) is what moves matter. But then, some physicists say that matter is "maya" or illusion, and that it is just condensed energy or ultimately information. However, this kind of thinking gives way to mysticism and takes physics out of the realm of the sciences in my opinion.

In one respect I would agree with Bohm in that a problem in English (and other languages too, I assume) is the *reification of nouns*, that is, making abstract nouns into concrete nouns. The familiar example I have used in this book is the *reification of space and time* and combining these concepts into spacetime. Einstein reified these two abstract concepts and made them into concrete, tangible entities. Thus spacetime, now made concrete, can be bent, curved, warped by mass and energy and can, in turn, affect matter and energy as it creates warps in the fabric of spacetime which cause smaller masses to move around larger masses in elliptical orbits and even affect the path of light. Einstein thus makes space and time active players in physical processes and the equivalents of mass and energy in the dynamic processes of the cosmos.

Michio Kaku (2015) refers to space as a "nothing", yet continues to speak in Einsteinian terms that space is an active force in the dynamo of the universe (Youtube Video). Let's analyze the word "nothing" in English to see how it is reified. The word "no-thing" is a compound word made up of "no" and "thing". "No" means negation or absence and "thing" is a noun that indicates a material substance (or abstraction). So "no-thing" is the absence of something – in a word "vacuum". So how can nothing be a force or an active player in moving smaller masses into orbits around larger massive bodies? I concur with Tesla who said that he could not endorse a theory built upon this fallacious concept of nothing affecting the motion of something. As previously stated, "spacetime" may be a metaphor that provides a prediction of how mass and energy interact and change in space through time, but I think it is a misnomer at best.

Hence I disagree with Bohm that English is not suited to convey the true nature of the cosmos. English is not totally a noun-based language that separates actor and action in my view, and is well-balanced between nouns and verbs, actions and agents. Its structure is a template that forces the speaker or writer to a complete thought - to answer the basic questions of "who, what, when, where and how" because its design forces the speaker to specify the agent of change (subject), the action of change (verb) and the object of change (direct object), complete with modifiers and prepositions to show relationships. Again, the fault I have found with English speakers is the *reification* of abstract nouns as indicated, and I assume that this is a tendency in other languages as well.

Despite its weaknesses and Bohm's idea that the truth of the cosmos transcends language, language is our primary vehicle for thinking and communicating our ideas as humans. We humans can readily see, relative to other animals, what an advantage language has conveyed to us in creating culture, technology and controlling the environment. We know that qualitative and quantitative language (math) are imperfect representations of nature which science tries to describe and explain - despite the fact that some physicists and mathematicians, like the Pythagoreans, think that math is somehow intrinsic or innate in nature.

Capra's Response to Criticism by Fellow Physicists

Capra's response to those physicists who criticize him for what they see as superficial similarities between Modern Physics and mysticism is that he is not alone in seeing a deep connection in mystical thought. At least two of the founders of quantum mechanics, namely Heisenberg and Bohr, believed in this intimate connection and acted upon it.

He (Heisenberg) said that he was well aware of these parallels. While he was working on quantum theory he went to India to lecture and was a guest of Tagore. He talked a lot with Tagore about Indian philosophy. Heisenberg told me that these talks had helped him a lot with his work in physics, because they showed him that all these new ideas in quantum physics were in fact not all that crazy. He realized there was, in fact, a whole culture that subscribed to very similar ideas. Niels Bohr had a similar experience when he went to China (Weber 1982: pp. 217–218).

Furthermore, Bohr made any number of statements that indicated that he perceived quantum physics as a mystical venture. When Einstein argued against the idea that quantum phenomena are random and without causation, Bohr responded: "Mr. Einstein, you're being too logical." And, when reviewing a quantum theory proposed by a colleague, he said: "Your theory is crazy, but it's not crazy enough to be true" (Brainy Quotes).

Piaget, Mysticism and Pre-logical Stage

Mysticism seems to be a return to pre-logical thinking that child psychologist, Jean Piaget described. In pre-logical stage of development, the child will believe her senses and perceptions and cannot apply logic to determine that her senses have deceived her. One best-selling sacred text says that "unless you become as a little child, you cannot enter the kingdom of heaven." Interpreted in the language of Piaget, Ornstein, Bohm and Capra, one might say that unless you revert to child-like thinking and use your right hemisphere to break out of the linear logic of the left hemisphere, you cannot enter the kingdom of modern physics and apprehend the secrets of mystical oneness.

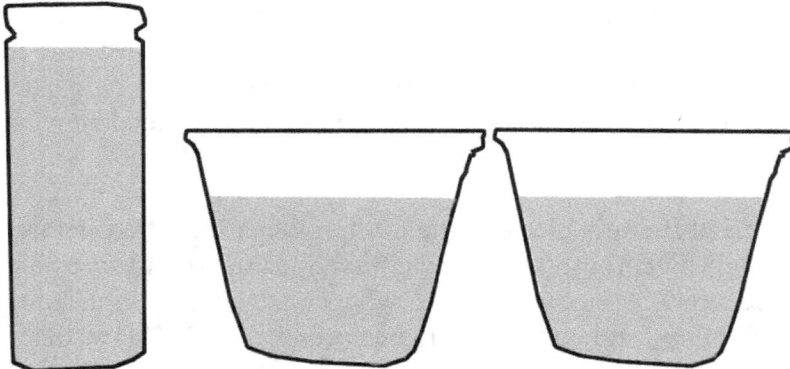

Conservation of Quantity Test: Both short glasses have the same amount of liquid. When the liquid from one of the short glasses is poured into a taller, thinner glass, a child that cannot *conserve* will say that the taller glass has more liquid than the shorter glass. This test can be given several times and the child in a pre-logical stage will continue to insist that the taller glass has more water. In other words, the child believes his senses and is not able to use logic to know when his senses are deceiving him. This pre-logical thought sounds eerily similar to some of the Zen koans that are designed to break the initiate out of the logical realm of thought into a transcendental realm which is considered to be higher than rational thought. In other words, the initiate must unlearn some of the things he has learned in the process of maturation and return to a child-like level of thought. The Zen koan of "What is the sound of one hand clapping?" seems to be a pre-logical type question. Similarly, in modern physics, the student is taught that she must give up the biases she has learned from interacting with slow, macro matter because fast and smaller matter (at the level of atoms) does not behave the same way as slow, macro matter. For example, the idea that a particle can be in two places at one time violates our sense of the conservation of an object's occupying a single position in space at a single point in time. And, the notion that a new universe splits off to accommodate each probability in a quantum superposition similarly violates our sense of the conservation of mass and energy. Furthermore, the Copenhagen interpretation of quantum mechanics indicates that reality does not exist until we observe it. In this view, our observation actualizes potentials and brings them into existence. This is very similar to a child in the pre-logical stage who can see the experimenter hiding something that he wants - like a ball - but does not apparently know it continues to exist and will not search for it. When it is out of sight of the infant, it is out of mind and no longer exists for him. Later in development, the child will search for the hidden object, and, in Piaget's words, she has mastered the logic of object permanence.

As a child learns perspective and angles, s/he becomes aware that a stick appears longer when looking at it from a right angle than looking at it from an angle other than a perpendicular one. However, after some interaction with objects in the real world, s/he learns that the stick conserves its length and that there is a superior way of looking at it in order to perceive its true length. Later, when the child matures into adolescence or young adulthood and takes a course in physics, s/he might be taught the Lorentz contraction in which a stick or rod traveling at a high rate of speed would appear shorter and that this would not be an optical illusion. From a classical perspective the student might reason that a rod or stick would not gain or lose any atoms while travelling very fast, but that acceleration could cause shortening of the object because the force that is propelling it could cause it to compress atoms, removing some of the space between them. But the student would be told that this is a common-sense answer and not the real answer. The real answer is that the unique properties of light interacting with massive bodies creates this effect and that the effect is real, not smoke and mirrors.

Likewise, a doctrine in relativity is that there is no *preferred frame of reference* – that all points of view are equally valid and if each perspective renders a different perception, each perception is true in its respective frame. That is to say, that the truth is relative and there is no absolute, preferred perspective. Hence if one twin is at inertial rest and the other is in inertial motion, each one perceives the other in motion since, without a background, they cannot tell who is moving and who isn't. Therefore, each one sees the other's time slowing down relative to his own time, and each twin's is correct. This conundrum leads to the twin paradox which will be discussed later. This thought experiment seems to violate the Piaget's conservation of quantity principle. As we mature from the pre-logical stage, we learn that some points of view yield a more holistic perspective than others, and we imagine ourselves into that more comprehensive perspective.

Piaget also spoke of an animistic stage that children go through in their cognitive maturation. In this stage, a child believes that there is no distinction between the animate and the inanimate and everything is alive and conscious. Furthermore, children may believe that their thoughts and desires affect the material world, and they typically believe the universe revolves around them in an egocentric (and geocentric) way. The child may say: "the sun and moon follow me because it is in my sky no matter where I go." When I was first reading Piaget, my daughter was very young. We had been on a rather long trip and were returning at night on a full moon. When we started our journey home, we noted the full moon, and when we arrived, my daughter said: "Daddy, look, the moon followed us all the way home." This incident increased my belief in Piaget. Moreover, in this stage the child believes that everything has life and consciousness. For example, the sidewalk and other inanimate objects have feelings.

In anthropology, we see that primal cultures tend to see the world as animistic. Anthropologist, Sir Edward Tylor, believed that primitive animism is the basis of all religion. In animistic thinking, everything has life and consciousness including inanimate objects such as stones and animate entities including plants and animals. We also see that some of Bohm's theories embrace animism. Bohm asserted that consciousness is present in varying degrees in all forms of matter including the electron. Like primal people, he believed that even a rock in some way is alive and conscious and "intelligence is present not only in all matter, but in energy, space, time, and the fabric of the entire universe (Talbot 1991: p. 50)." Perhaps, Bohm meant that everything contains information – even inanimate matter – and that information is a kind of consciousness. It is interesting that Bohm studied Piaget's theories and saw the cognitive development of children as a manifestation of the implicate order. He held that Piaget's studies show that young children learn about time and space because they have a "hard-wired" understanding of movement as part of the implicate order (Bohm 1980). This line of thinking sounds very Greek in the sense that the mind is endowed with innate ideas, or at least innate structures. This kind of animistic thinking is eloquently expressed in Zukav's Dancing Wu Li Masters who contend that all reality is "organic", that is, everything is alive and each part is mystically connected to every other part. Perhaps, even the idea of a separate, independent part is illusion and that all reality is an indivisible whole.

At any rate, Bohm's thought process gives a kind of scientific explanation for the primal concept of animism. The ultimate animism is the Gaya principle which is the belief that the earth itself is a living, breathing organism. And, incidentally, some believe that we humans are the cancer on this living organism.

The **anthropic principle** partakes of a kind of animistic thought as well in that it asserts that the universe was created in such a way as to make life inevitable or with the strong potential for life. There are varying degrees of the anthropic principle, but the participatory anthropic principle championed by John Wheeler is the most animistic. Below are the four forms of the anthropic principle simplified.

Weak: the universe was probably made for man.
Strong: the universe was definitely made for man.
Participatory: the universe was made by man.
Final: the universe will be occupied by man.

The next chapter will deal with the Anthropic Principle and its implications for consciousness as a factor in physics theory.

> Eastern mysticism or any brand of mysticism has its place in philosophy, religion, and meditation, but it is unscientific. In the early 1900's, mysticism got mixed in with science in the quantum revolution in physics thinking. Quantum mysticism was said to be counter-intuitive and could not be understood with common sense logic or rationality. Now, some physicists are trying to divorce quantum theory from mysticism and rewrite history by saying that it never had anything to do with mysticism despite the many Zen-like koans inherent in the theory. Neils Bohr studied Eastern Mysticism and applied it to quantum mechanics. When

Einstein objected to the theory of quantum entanglement which indicates that signals can pass between entangled particles instantaneously (that is, faster-than-light), Bohr answered: "Mr. Einstein, you're being too logical." – thus giving a voice to mysticism.

CHAPTER 2: THE ROLE OF CONSCIOUSNESS IN PHYSICS THEORY

It almost goes without saying that the goal of physics theory is to become more conscious of the universe and the laws that govern its operation. Isaac Newton thought that the study of physics was the study of the mind of God. Thus physics, to Newton, was a path to a kind of God consciousness, that is, to see the universe through the eyes of God. The question becomes: Is our consciousness independent of reality having no direct effect on it, or does our consciousness affect reality, and are we humans a part of the creative process? Is there an objective reality out there that is independent of our consciousness, or is all reality subjective? When quantum physicists first stated that the observation of the experimenter determined the outcome of an experiment, consciousness was let in through the laboratory door and into the universe - theoretically. Since that time, many physicists have given up the idea of an objective reality that is independent of our perceptions. This concept that one's consciousness affects reality is a very primal one as we see with a Navaho shaman who believes he chants the world into existence each morning. Talbot (1991) sees this trend as the shamanizing of physics, and the anthropic principle is one expression of this idea.

The Anthropic Principle

While some see the anthropic principle as non-theological, other see is as confirming theology, particularly, those of the Intelligent Design persuasion. Intelligent Design theorists say that when theorists speak of the universe as being designed in such a way that humans would inevitably come into existence, then they are implying a "designer". Even if one uses other words, the implication of a designer is always present. For example, if you say: "The universe was structured in such a way as to make it possible for humans to evolve", you are implying a supernatural finger in the pie. Such a statement would inevitably evoke the question of "who" structured it in such a way with this end in mind. It is difficult to imagine an impersonal process having any goal or aim if there is no mind or consciousness associated with it. Teleological statements such as the anthropic principles imply a mind - not an impersonal, random process. This non-theistic anthropic principle is reminiscent of Einstein's many references to God in his theoretical ruminations, for example, "God does not play dice." Yet when questioned about his religious beliefs, Einstein said he didn't believe in a personal god but did believe in an impersonal god. How the idea of an impersonal god can reside under one skull without serious cognitive dissonance is difficult to understand. God, by definition, is a personal being who has consciousness and will - much like human beings – except God is conceived as a super human being. If the universe is impersonal, that would mean that it came into being unconsciously and without a plan. It would have evolved from the interaction of matter and energy according to impersonal laws of nature with no end goal or intention (teleology). The idea of an "impersonal god" is a self-contradiction that can only be apprehended through mysticism where contradictions can co-exist in the same concept.

Biological evolution of the Darwinian variety indicates that the whole process of the evolution of life leading to man was random and non-teleological. Without a mind operating in nature, there is no aim or goal to be achieved in the future – the present form of the world must be explained in terms of the past, not the future. Here are some contrasts between the anthropic principle and Darwinian evolution.

Anthropic principle vs. Darwinian evolution

Teleological vs. randomness
Future determinism vs. past determinism
Intended vs. unintended
Directional vs. non-directional
Universe adapted to man vs. man adapted to universe

Surely most Darwinian biologists would see the anthropic principle as mystical, rather than based on natural law. The essence of biological evolution is the idea that random mutations are selected for by an unforgiving environment. Those who happen to win the cosmic lottery of having the most adaptive traits survive and reproduce, passing on their those traits to the next generation which must meet the imperative of survival or extinction. Darwin might say that life on earth was not predestined or inevitable – life has no destiny, it only has a past and present with no guarantee for the future.

The participatory anthropic principle, espoused by physicists such as John Wheeler, endows man with God-like powers to create the universe retroactively with the mind. Wheeler believed that we humans had not only created the universe with our thoughts but had generated its history by sending our thoughts back in time. Again, as an anthropologist, I see a close connection between the participatory anthropic principle and the Navaho shaman who rises early each morning to chant the sun up and to chant the world into existence each day (Mabury-Lewis 1992).

Now for those materialist philosophers who believe in evolution, the universe, the earth, and the living organisms came into being long before humans came into existence. The universe is said to be somewhere between 13 and 15 billion years old, and modern humans are believed to have come into existence some 100 thousand years ago. But to Wheeler, that chronology is no problem because, according to his interpretation of quantum reality, our consciousness can somehow create things in the past. This theory seems to this author to be a return to mysticism. How can something created before humans came into being, then be recreated as though it were not there in the first place?

Moreover that same John Wheeler, who believed that we humans had created the universe with our consciousness, called for the expulsion of parapsychologist from the American Association for the Advancement of Science (AAAS), condemning their discipline as pseudoscience because, as he said, there was no evidence for the "psi" factor. However, it would seem easier to me to believe that a man like Uri Geller could bend a spoon with his mind than to believe that humans could create the whole universe and its history with their minds. Years ago, I did some research in parapsychology under J.B. Rhyne whose research center had been moved off the Duke University campus. At the time, there was a physicist who was attempting to find the energy that carries the psi factor – a fifth force if you will. Doing this kind of research presupposed that he, like Bohm, believed in the existence of psychic phenomena. So, not all physicists renounce the connection between physics and parapsychology. This connection seems an inescapable conclusion in the quantum theories that posit that the observer's consciousness determines the outcome of an experiment.

Moreover, this same John Wheeler, after supporting Everett and DeWitt in their development of the Many Worlds Theory, which posits that the universe splits to actualize each quantum probability, later rejected the theory because "it carries too great a load of metaphysical baggage" (Gribbin 1984: Kindle Book 84%). One might reasonably ask the question as to whether the Many Worlds Theory carries more metaphysical baggage than Wheeler's Participatory Anthropic Principle. John Wheeler was certainly a man of many contradictions.

Now Gribbin (2009), in his book, *In Search of the Multiverse*, says that the anthropic principle is "the argument that the universe we see has to be the way it is, or we would not be here to see it. This does not mean necessarily that the universe has been 'designed' with us in mind" (p. 206). This may be a statement of the weak anthropic principle, but it is certainly not how Davies and Deutsch describe the strong anthropic principle in a video by that name (2006). Perhaps we could paraphrase Davie and Deutsch and say that given the laws of nature, it was inevitable that we humans would evolve. Of course, this gives much ammunition for the Intelligent Design theory that the universe was intentional and not accidental. What Gribbin states here is circular reasoning and tautology. In its simplest terms, he is saying if it were not for carbon, we would not be carbon-based beings. This line of reasoning seems to get the cart before the horse and begs the question as to whether some other life form might have evolved based on another chemistry or physical laws somewhere in our own universe or in another (if indeed other universes exist). Even on our own planet, we find tube worms attached to the ocean floor near vents spewing hydrogen sulfide. The tube worms are able to metabolize hydrogen sulfide boiling from those vents, which is extremely poisonous to human beings and other animals. So even here on the earth, we see a somewhat different chemistry among various creatures living in very different niches of the environment. Perhaps there is intelligent life somewhere else in the universe founded upon a very different chemistry than the carbon-nitrogen based life on the earth. At any rate, certainly the strong anthropic principle is more anthropocentric in stating that human consciousness creates the universe and its history going back in time.

The Parapsychology and Physics Connection

Physicists opened up Pandora's Box when they said that the observer's consciousness determines the outcome of a quantum experiment and that time is not a fixed entity but can be sped up or slowed down. The opening of that proverbial box provides shaman and psychics with scientific validation for what they believed all along. The purpose of this book is not to try to prove or disprove paranormal phenomena, but to show how the implications of modern physics inevitably lead to mystical conclusions as some physicists have already acknowledged but which others deny vehemently.

Parapsychology is the study of ESP (Extra Sensory Perception) and PK (Psychokinesis, sometimes referred to as Telekinesis). The following is a breakdown of these types of alleged paranormal phenomena.

ESP: Extra Sensory Perception involves perceiving something without the mediation of known physical senses such as sight and hearing.
1~**Mental Telepathy**: ESP involving reading another person's mind without known sensory mechanisms such as language or body language

2~**Clairvoyance**: ESP involving perceiving a distant event or object without known sensory information
3~**Precognition**: ESP involving predicting future events – future-telling

PK (Psychokinesis) moving objects or affecting physical phenomena with the mind at a distance without physical contact or mechanical means. PK might be called spooky action at a distance.

Modern physics has inadvertently given support to the mystical types in society who feel that their spiritual beliefs have been confirmed by science. For example, some psychics claim that their ability to foresee the future is a result of information traveling backward in time to them from the future to the present – derived from quantum theories that speak of information traveling backward in time from the future. Or, some psychics might believe that they have the ability to time travel to the future through the powers of the psyche, and correspondingly, some physicists believe that the laws of nature do not forbid time travel.

While some physicists eschew what they see as mystical exploitation of their discipline, others embrace the seemingly obvious implications of the new physics with enthusiasm. David Bohn, mentioned previously, who famously coined the term 'the seamless whole" was one of those physicists. I admire Bohm for his audacity in following the implications of physical theory to their logical conclusions without waffling or wavering. Other physicists, with more of a materialist philosophy, have forcefully denounced the connection that some have made between seemingly supernatural, non-physical phenomena and the findings of the new physics. The role of consciousness in determining the outcome of physics experiments is one of those claims that fuels the mystical connection. It seems obvious that if observation collapses a wave function and determines the outcome of a quantum experiment (whether the cat is dead or alive), then consciousness can affect material reality – parapsychologists call it psychokinesis (PK) as noted above. Bohm believed that physics confirms the effect of mind on matter and embraced Uri Geller, the Israeli psychic, who claimed that he could bend spoons with his mind. Bohm was warned that his association with Geller could undermine his credibility as a physicist. Such warnings were based more on sociology than physics theory – because the field of physics has respectability as a science, and parapsychology has little respect.

However, I would like to raise a bold question as to whether there is more evidence for "psi" factor than there is for the extra dimensions of string theory or the multiverse in the many worlds theory. Recall, that Wheeler criticized parapsychology for lack of evidence for the psi factor. As indicated previously, as a graduate student, I became fascinated with "scientific" psychic research and did some study under J.B. Rhyne who was then located in Durham near the Duke University campus. I understand that Duke University, particularly its psychology department, did not want to be associated with this research because many considered it pseudo-science. At any rate, I had wondered that if psychic phenomenon exists, would it be more prevalent in primal cultures where there is a strong belief in it and where lack of scientific rationality would inhibit primal mental processes less. In my studies with J.B. Rhyne, however, I concluded that if psi exists, it is a very weak factor. The best psychics could average only 7 out of 25 (2 above chance) on the Zener card test. Nevertheless, in a large number of runs, this small deviation from chance could be statistically significant. Since I was expecting a much higher deviation from chance, I became somewhat disillusioned with parapsychology and pursued other studies. However, Rhyne (1947) shows how significant a small deviation from chance can be in a large number of runs:

The best individual performer went through the deck of ESP cards well over 700 times during the first three years of work covered by my first report...This man averaged about 8 hits out of 25 tries...To express the odds against averaging a score of 8 or better by chance alone for more than 700 runs would require a paragraph of figures...(p. 35)

Now, back to my question, is there as much evidence to support the more exotic theories of modern physics such as string theory or the many-worlds theory as there is the "psi" factor? I don't think so, but let an expert speak to this. At least the American Association for the Advancement of Science (AAAS) recognized the study of paranormal phenomena, including ESP, to be a valid scientific investigation, and John Wheeler's attempt to get parapsychologists expelled from this organization failed. It is tempting to make a connection between the alleged instantaneous communication between quantum particles at a distance ("spooky action at a distance" as Einstein called it) and the alleged instantaneous perception of an event at a distance by psychics. Here again, psychics find support for their beliefs in the exotic theories of physics.

To determine whether ESP is greater among primal peoples who may have a stronger belief in it, the Roses (husband and wife team) tested Australian Aborigines. They recorded 16, 625 guesses, chance would have given 3,325 hits, but they recorded 3,870 correct hits (Rose 1956: p. 227) which the Roses thought was evidence that the Australian Aborigines had ESP abilities. In a questionnaire which was submitted to the natives, most of the Roses' subjects answered "Yes" to the question, "Would you know if a relative some distance away died, had an accident, or was seriously ill? Rose cites some anecdotal cases which he believes was indicative of ESP.

In primal cultures and modern cultures, there is a belief that entering into an altered state of consciousness may enhance paranormal abilities. In tribal societies, a shaman might induce a trance similar to hypnosis or take a psychoactive drug to enter another realm where normal abilities are transcended. In India, yoga meditation practices are thought to enhance one's receptivity to information through a channel of the mind other than the five major senses. And in modern, technological societies, biofeedback machines have become a kind of shortcut to nirvana where unusual control of autonomic functions of the body has been achieved. Anthropologists have reported some rather unusual personal experiences with shaman in the field. For example, a Mexican curandero in a remote village gave Gordon Wasson information about his son, telling him that he was not in Boston, as the Wassons believed, but in New York, and he would soon be in military service which turned out to be true (Wasson and Wasson 1957: 2, 264-65).

Apparently Bohm is not the only physicist who believed in paranormal phenomena. At the 1987 annual meeting for the Association for the Study of Dreams, physicist Fred Alan Wolf delivered a speech in which he asserted that the holographic model explains lucid dreams (unusually vivid dreams in which the dreamer realizes that he or she is dreaming). Wolf believes such dreams are actually visits to parallel realities, and the holographic model will ultimately allow us to develop a "physics of consciousness" which will ultimately enable us to explore other dimensional levels of existence (Talbot 1991: p. 3).

Psychokinesis (PK) and Modern Physics

Bohr said that the mind is melded to physical reality and there is no objective reality that is separate from the mind. Yet, some physicists say that this mind-reality unity has nothing to do with psychic phenomena. This seems to be a case of not owning the implications of one's theory.

The advent of modern computers, which can generate random numbers, has enabled a statistical analysis of PK and ESP similar to the statistical analysis of probabilities in quantum experiments such as Bell's inequalities. This research has come to be known as the **PK Effect on Random Number Generators**. Anyone who has played the lottery and allowed a computer to select a random number is familiar with a random numbers generator, and I am sure there are many lottery participants who try to use psychokinesis to cause their set of random numbers picked by the computer to correspond to the winning random numbers.

Like the would-be lottery player who tries to determine the numbers drawn with her mind, in the typical Random Numbers Generators experiment, a subject attempts to mentally change the distribution of random numbers that a computer generates. The typical experimental design is similar to getting more "heads" than "tails" while flipping a coin or getting a particular number in a roll of dice. The Parapsychological Association (2011) asserts the following regarding these experiments.

A meta-analysis of the data, published in 1989, examined 800 experiments by more than 60 researchers over the preceding 30 years. The effect size was found to be very small, but remarkably consistent, resulting in an overall statistical deviation of approximately 15 standard errors from a chance effect. The probability that the observed effect was actually zero (i.e., no psi) was less than one part in a trillion, verifying that human consciousness can indeed affect the behavior of a random physical system…However, the apparent effect of focused mass consciousness on a world-wide network of RNGs (see the Global Consciousness Project) suggests that at least some of the time, there is an element of mind-matter interaction.

However, it would seem to be difficult to ascertain whether these results should be attributed to ESP or PK. If precognition (future telling) is a reality, it could be that the subject is using precognition to discern what numbers will come up, rather than causing numbers to appear with psychokinesis. However, since the object is to cause a distribution of numbers to deviate from chance expectation, then PK could be distinguished from precognition if such a deviation is observed.

Let us further examine the troubled, on-and-off relationship between physics and the paranormal. Mention has already been made of the Copenhagen interpretation which indicates that the observation or the experimenter's consciousness determines the outcome of a quantum experiment. If the experimenter observes a wave, it becomes a particle and observation determines the outcome of the experiment. The boldest theory in physics which supports psychokinesis is the participatory anthropic principle developed by physicist John Wheeler. This theory, based on the delayed choice quantum experiment, is that human consciousness has created the universe. Wheeler seems to be saying that the universe was in a state of quantum uncertainty created by the Big Bang, i.e., the universe existed and didn't exist until humans evolved consciousness and observed it, thus making the universal wave collapse into a definite particulate state – even as an experimenter's observation of a probability wave forces it to become a real particle with a definite place in space and time. If Wheeler is correct, then only part of the universe has been actualized since the whole universe has not yet been observed by humans. Despite this rather definite support for beliefs in psychokinesis from physics, there are other physicists who take a hard materialist point of view and attempt to debunk PK and ESP, and some have tried to distance the observer effect in quantum physics from PK. Philosopher and physicist Mario Bunge (2001) has written that

Psychokinesis, or PK, violates the principle that mind cannot act directly on matter. (If it did, no experimenter could trust his readings of measuring instruments.) It also violates the principles of conservation of energy and momentum. The claim that quantum mechanics allows for the possibility of mental power influencing randomizers — an alleged case of micro-PK — is ludicrous since that theory respects the said conservation principles, and it deals exclusively with physical things. (p. 196).

Bunge's contention that the connection made between quantum physics and PK is ludicrous seems like a rationalization and denial to yours truly. Certainly, the Many Worlds Interpretation does not respect conservation principles with the universe splitting for every quantum possibility. Furthermore, if the cat is dead *and* live in the Copenhagen interpretation and the consciousness of the physicists makes it one or the other – dead *or* alive - then what happened to that other state which is not manifest. If the cat is made to be dead by observation, then what happened to the alive state? Destruction of the opposite state also seems to violate conservation laws. The idea that quantum physics deals only with physical things is also fallacious. Consciousness, although connected to the physical brain, is, in itself, non-physical. If consciousness affects the physical outcome of an experiment, then that would be the same as the mind affecting macro or micro reality. It appears that Bunge is making a distinction without a difference. Bunge doesn't seem to know about or account for John Wheeler's participatory anthropic principle which posits that human consciousness has created the universe. The claims of parapsychologists are quite tame compared to Wheeler's extravagant claim.

Some parapsychologists have proposed a fifth force in nature (the psi force) in addition to the four known forces of electromagnetism, gravity, the strong and weak force. Physicist John Taylor *(1980)*, who has investigated parapsychological claims, has written that

An unknown fifth force causing psychokinesis would have to transmit a great deal of energy. The energy would have to overcome the electromagnetic forces binding the atoms together, because the atoms would need to respond more strongly to the fifth force than to electric forces. Such an additional force between atoms should therefore exist all the time and not during only alleged paranormal occurrences. Taylor contends that there is no scientific trace of such a force in physics, down to many orders of magnitude; thus, if a scientific viewpoint is to be preserved, the idea of any fifth force must be discarded. Taylor concluded that there is no possible physical mechanism for psychokinesis, and it is in complete contradiction to established science (pp.27-30).

Other physicists have denounced psychokinesis on the grounds that it violates the inverse square law of the diminution of energy as it spreads out in space. Again, a double standard emerges, allowing physicists to claim that *entangled particles communicate with each other instantaneously over great distances*. In this claim for entangled particles, no inverse square law is observed, since, theoretically, an entangled particle in the Andromeda galaxy could instantaneously exchange information with its partner in the Milky Way even though the two galaxies are said to be 2 million light years apart. Where is the inverse square diminution of that energy which carries the information between entangled particles? Perhaps physicists would also want to invoke a fifth force to explain quantum entanglement since the information between entangled particles is instantaneous and shows no fall off with distance. Now since the maximum speed for all known forces is the speed of light, it seems that a fifth force is indeed necessary to explain the instant communication between quantum particles at any distance. Moreover, there are physicists who have hypothesized that the brain is rather like a quantum computer in which information is carried by entangled quantum particles in microtubules in brain proteins (Hameroff 2008). If this is so, then perhaps the psi energy is rather like the quantum energy that allows instant communication between entangled particles and would therefore qualify as a fifth force.

Despite all these denials by physicists, there seems to be no escape from the quantum physics/psychic connection. If it is unscientific for psychics to claim they can perceive things at a distance or move objects with the mind, then we must discard the double standard that indicates that the observer's consciousness determines the outcome of quantum experiments and that the mind created the universe. If there is no fifth force for psychics, then there is no fifth force for quantum physicists either. We need to drop the hypocrisy that says that "consciousness affecting reality" is respectable science for physics, but disreputable pseudoscience for parapsychology. These debunkers should be aware that they are also condemning their fellow physicists who have embraced the psi-physics connection. Some physicists, such as David Bohm, have had integrity enough to embrace this connection.

> **Degree to which consciousness affects reality… There are three hypothetical levels:**
> 1>Consciousness creates reality totally from whole cloth: Perhaps John Wheeler's participatory anthropic principle would be an example of this.
> 2>Consciousness shapes reality but doesn't totally create it: An example of this would be the Copenhagen interpretation in which the observer's consciousness collapses a wave and causes one of a number of possibilities to become physically real. In Schrodinger's cat box thought experiment, it is not the observer's intent for the cat to be dead or alive that causes the cat to be one way or the other. It is the observation that takes the cat out of limbo and makes it either dead or alive, not dead and alive.
> 3>Consciousness does not directly affect reality. Consciousness only affects reality if some instrument that physically affects reality is used to probe it. In Heisenberg's uncertainty principle, the observation of a particle's position by hitting it with a light makes it impossible to know the particle's momentum or speed since hitting it with light changes its position and momentum.

A Proposed Experiment to Show Consciousness Does Not Affect Quantum Entanglement
Since some physicists have become uncomfortable with the injection of consciousness into quantum phenomena, these physicists want to eject consciousness from this theory to avoid the stigma of mysticism and thus return physics to hard materialism and rational empiricism. A proposed experiment by Lucien Hardy involves the following set up (Ananthaswamy (2017).

1~Setting up a quantum entanglement type experiment according to Bell's inequalities in which entangled particles are sent to different detectors some distance apart.
2~A random numbers generator determines some quantum property such as spin of one of the particles. The entangled particle will show the same or complementary property at the other detector, suggesting instant communication between entangled particles.
3~Experiments of this type have shown a statistically significant correlation between entangled particles at a distance in terms of spin and other quantum properties. For example, if one particle is made to be spin up at location A, the other will be spin down at location B, no matter how far apart they are.
3~To inject mental interference with the particles' communication with each other, several humans would be hooked up to EEG headsets and the signals from their brain activity would be used to determine the state of the particle rather than a random numbers generator.
4~If the EEG signals prevented the entangled particles from correlating their behavior at a statistically significant level as is measured without human conscious interference, then consciousness would appear to affect physical reality in overriding entanglement.
5~However, if the entangled particles continued to show the same high correlations as they did without conscious interference, the result would indicate that human consciousness does not affect quantum behavior.
6~Another suggestion is that, rather than use EEG signals, get subjects to try to control the quantum state of the particle on one end or the other and see if that prevents the high correlation between entangled particles.

This proposed experiment is very similar to the random numbers generator experiments in which research subjects try to determine the sequence of numbers generated by a computer. However, computer-generated random numbers are said to be Pseudo-Random Number Generators. To get true random numbers, computers must be connected to some outside natural phenomenon which gives true randomness. According to Random.org, a really good physical phenomenon to use is a radioactive source. The points in time at which a radioactive source decays are completely unpredictable, and they can quite easily be detected and fed into a computer. Whatever the case regarding how random numbers are generated, the small, but significant, deviations from chance derived from random numbers experiments have *not* been accepted by physicists who have critiqued these experiments. But, perhaps, the notion of the mind overriding quantum entanglement correlations would be more impressive to the critics of the random numbers generator experiments.

> The Copenhagen interpretation is somewhat different from PK although they share the notion that consciousness affects reality. The Copenhagen interpretation is that consciousness actualizes a reality that is already potentially there but in multiple states. The observation of the experimenter forces the superposition to become one state or the other, i.e., the cat to be either dead or alive, whether the experimenter wants the cat to be alive or not. PK, on the other hand, is making physical reality the way one wants it to be, that is, making the cat live rather than die if the psychic loves cats. Both cases, however, involve mind influencing physical reality without physical contact – both involve "spooky action at a distance."

Near Death Experiences (NDEs), Soul Travel and Modern Physics

If altered states of consciousness such as meditation, hypnosis, ecstasy and trance connect one's perceptions to sensory channels that are yet unknown to science, then perhaps the ultimate altered state is a near death experience or NDE. To some, NDEs offer even more – they strongly suggest to some people (including some in the medical profession) that the mind or spirit can continue to exist and be conscious apart from the brain and body. Again the question arises, does physics say or imply anything about the veracity of this ancient religious belief? What do particle colliders and relativity tell us about these issues relating to consciousness? Consciousness has already entered the collective mind of physics in the quantum experiments whose outcomes are dependent upon the observation of the experimenter – in other words, the outcome depends on the experimenter's consciousness. And John Wheeler would go so far as to say that the universe and its history depends on our consciousness. Are we still within the scientific domain in these edge issues?

Anthropologist, Sir Edward Tylor, speculated on the origins of religion in his seminal work, *Primitive Culture* (1877). Tylor argued that the belief in soul comes from universal human experiences which seem to take place outside the body – experiences such as dreaming, trance states, drug intoxication and perhaps near-death experiences. In all these altered states of consciousness, the person perceives that her spirit or consciousness separates from the physical body and is free to travel to different places and even different dimensions. When the person awakes or comes back to normal consciousness, she finds that her body is where she laid down initially, and she interprets this return as the reuniting of soul and body. From this notion of disembodied spirit comes the various forms of religion as it evolves into more convoluted forms in more complex societies. The underlying belief seems to be that the disembodied spirit is constantly seeking a body or physical object to inhabit. Thus, spirit or anima becomes projected on all objects in nature. First there is animism where everything including inanimate objects has spirit, then comes totemism where animals and plants have consciousness and spirit, followed by polytheism and super human type gods that can be represented with idols or statues, and finally monotheism where there is one god that is a pure spiritual being that cannot be represented by a material object.

However, modern scientific psychology, particularly behaviorism, embraces materialism and generally denies the separate existence of spirit and body. The modern view is that the mind and consciousness are functions of the brain, and when the brain is altered, physically or chemically, consciousness is altered, and when the brain dies, consciousness dies with it.

Now we find some scientists, especially medical science practitioners, who have come to believe that the spirit may survive physical death and continue an independent existence. Some have been convinced of this independent existence of the soul from studying patients who have had dramatic Near-Death Experiences. The typical NDE experience in hospital settings is that as the patient is nearing death, her spirit rises above the operating table where she can witness what the doctors are doing to her body and can hear their conversations. Sometimes the NDEer's spirit allegedly roams about the hospital and perceives objects that would be out of her view on the operating table. Sometimes the NDEer experiences psychic phenomena and perceives something that no known sense could have informed her of. Such was the case of Dr. George Rodonaia, a former Soviet scientist and dissident who made his way to the US. While standing on a sidewalk, he was struck by a car. He was pronounced dead but somehow returned to life. While in the near death (or death state), he had a vision that the young daughter of a friend had an undiagnosed hip abnormality, and this perception was later confirmed after Rodonaia returned to normal consciousness, and the daughter's orthopedic problem was subsequently treated (Shockey 2013 documentary).

Several doctors have teamed up to study frequent reports of NDE's experienced by their patients, and some physicians have reported having such experiences themselves. Dr. Sam Parnia is the principal investigator of these reports. The project is called The AWARE study (AWAreness during REsuscitation). AWARE is a multi-hospital clinical study of the brain and consciousness during cardiac arrest, including testing the validity of perceptions during the out-of-body phenomenon of near-death experiences (NDEs). The initial results, from the first four years of the study, were published in December 2015 in the medical journal *Resuscitation*.

Of the 2,060 cardiac arrests during the study, 140 patients survived and could be interviewed for the study. Of these, 101 patients had detailed interviews, which identified 9 patients who had an NDE. Of the 9 NDErs, two had detailed memories with awareness of the physical environment. One NDEer's experience was verified as accurate; the other was too ill for an in-depth interview. These two NDEs occurred in non-acute areas where no visual target was present, so further verification of visual awareness was not possible (2016: IANDS website).

The following is a summary of the NDE patient whose perceptions were deemed accurate.

1) During the NDE, the patient felt quite euphoric.
2) The patient heard an automated voice saying "Shock the patient, shock the patient."
3) The patient perceived that his spirit rose near the ceiling and looked down on his body, the nurse and another man, bald and "quite a chunky fella", who wore blue scrubs and a blue hat.
4) The patient could tell the man was bald because of where the hat was.

The next day, the patient recognized a bald man who attended him during the resuscitation. The medical record confirmed the use of an AED (Automated External Defibrillator) that would give the automated instructions the patient heard and the role that the identified man played during the resuscitation.

In the circle of my acquaintances, I have known three people (one woman, two men) who have had NDEs, and I have listened to their accounts. The two men were heart patients and the woman had respiratory problems. All of them had the typical experiences of the separation of their spirits from their bodies, going through a dark tunnel and seeing the light at the end of the tunnel, experiencing clear light, engaging in telepathic communication, etc. The woman indicated that she saw Jesus on a throne adorned with rare jewels, and she said she actually had a conversation with him in which he told her she must return to earth and complete her mission. One of the men, a minister, also indicated that he saw Jesus. In other NDEs, patients report supernatural healing of their bodies from a disease or injury that brought them to the point of death. For example, Rodonaia recovered from the injuries of being run over by a car and later became a Methodist minister.

In my reading of NDEs, I have found that many of the experiences involved concepts very similar to mysticism and modern physics especially in regard to time and space. For example, time takes on a different meaning and does not flow in the usual sense but exists in the eternal present. Some claim to have experienced the past, present and future in a kind of block time, and others say that time and space do not exist in the usual sense. Most report being pulled through a tunnel toward a clear light. This experience might be interpreted in physics terms as traveling through a spacetime portal (or wormhole) to a different dimension or perhaps to a different universe. The NDEer usually interprets arriving in the clear light as reaching a heavenly dimension of unconditional love and absolute bliss. Dr. Jeffrey Long (2010) has collected thousands of NDE cases and has performed statistical analysis to find the commonalities in these experiences. Here is what he found about the altered sense of time and space which are eerily similar to the concepts of time and space in relativity and quantum physics. The following is the case of a diver that typifies these altered perceptions.

When I first left my body I had my diving watch on. I took some very unscientific measurements of the distance I traveled by watching for features and measuring them by the second hand on my watch. Totally unscientific. But my conclusion was and has always been: I was measuring time in an altered time. The ground never moved in a linear fashion; the distances were erratic at best. The distances were always changing, sometime[s] repeating and then instantly becom[ing] longer or short[er] than the previous distance. Yet my watch was always ticking without change. My intuition and impression were that I was in a different time zone, one where my earth[l]y watch was of no use or inept at making any measurement or reflecting time. Also without mistake I would say this whole thing took an hour or more. It seemed to me that I was in the NDE for a very long time. But when I asked my diving partners how long had I been unconscious, they estimated five to ten minutes. Thus I had another reason to support why my diving watch didn't seem to measure the time in my NDE. It seemed as though I experienced so much in such a small length of earthly time. Where my soul had traveled to know nothing of time as we know [of] time passing on earth. Both time and space in earth stopped completely. Simultaneously, "the time and the space" on the other side was completely alive, evident[ial], and real. Yes, while I was in the light, I had ... [no] sense of time as I know it here on Earth. In other words, no sense of the serial nature of time ... past, present, or future. All times (past, present, and future) were experienced at every moment in time while I was in the light (p. 12, 13).

This case typifies the statistical profile of the altered sense of time in the large collection of experiences from thousands of people. These reports sound very much like time dilation and space contraction/expansion in Einstein's relativity. According to Einstein, for light, time goes to zero. Perhaps this could be interpreted as time ceasing to exist in the presence of a light being.

The NDERF survey asks, "Did you have any sense of altered space or time?" To this question the majority, 60.5 percent, answered "Yes." Another NDERF survey question focused only on an altered sense of time, asking, "Did time seem to speed up?" NDErs responded to this question with 33.9 percent selecting "Everything seemed to be happening all at once (p. 13)."

So what are the implications of modern physics that might explain these extraordinary phenomena? Hawking (2010) has claimed that "philosophy is dead" and that physics has the answers to life's ultimate questions. If consciousness can determine the outcome of an experiment as the Copenhagen interpretation claims or create the universe and its history as Wheeler said, or if the material world is an illusion, is it possible that the human spirit survives the death of the material body since the body may be an illusion after all?

Those of a materialist persuasion claim that NDEs are hallucinations caused by oxygen deprivation or neurons firing as they depolarize while the patient nears death. Others have claimed that NDEs (or the sensation of soul travel) can be induced by psychedelic substances such as ketamine or DMT, indicating that NDEs are a function of the brain in an altered state of consciousness. However, many NDErs have argued that what is experienced in an NDE is much more vivid that a dream. Furthermore, NDEs seem much more coherent and understandable than a dream or hallucination which often involves bizarre, disconnected or schizophrenic-type imagery and cognition. Many NDErs have a transformative experience that makes them better, more altruistic people.

Further arguments against the validity of NDEs include
1) Culture determines or influences the form of NDE (Christians may see Jesus, Buddhists may see Buddha). Thus it is seen as a subjective experience determined by one's culture and background.
2) Not everyone has an NDE - About 10% of heart attack patients experience an NDE.

As stated earlier, my purpose here is not to try to prove or disprove the validity of NDEs or other mystical experience, but to show that the implications of modern physics are similar to the beliefs in paranormal phenomena.

Psychedelics, Soul Travel and Near Death Experiences

As indicated above, many connections have been made by physicists and other professional scientists between modern physics and mysticism. One such professional is Dr. Karl Jansen, who is a psychiatrist that sees intimate connections between quantum theory and altered states of consciousness brought on by psychedelic drugs such as ketamine. These psychedelic drugs produce experiences closely resembling NDEs. Jansen began his research on the role of ketamine in psychotherapy as a classical scientist with a materialist mind set. In other words, he believed that psychedelics create hallucinations by changing the electro-chemistry of the brain and that these hallucinations have no reality outside the person's mind. Now that he has experienced a ketamine trip himself, Jansen (1999) has changed his mind and believes that ketamine and other psychedelics lead to another realm and provide evidence that the mind can exist independently of the brain. Jansen's argument goes like this.

Psychedelics open the doors of perception (as Huxley said) to a higher realm of reality that is quantum-like rather than classical or relativistic. In this quantum consciousness, space and time no longer limit human awareness to a fixed point in spacetime. Just as a quantum particle can be in more than one place at a time and entangled particles can communicate instantaneously across the universe, so human consciousness can span the universe and be aware of things beyond our limited sensory scope if the shackles on the mind are removed by ketamine or perhaps an NDE. Thus consciousness and communication are not limited to the speed of light as Einstein argued, so quantum consciousness is a realm above the relativity domain where simultaneity (the eternal present) can be achieved despite relativity's prohibition against simultaneity of distant events.

While it is indeed a quantum leap to generalize from the behavior of a photon (light particle) to the behavior of a human being - a macro object, Jansen insists that there is a convergence of the reality of the very small to the reality of the very large with the help of the psychedelic ketamine which can transport the human mind (without the encumbrance of the body) to a higher realm governed by quantum, not classical, laws. Jansen (1999) holds to this belief despite the fact that quantum physicists argue that the quantum realm obeys different laws from the macro classical realm in which we live. In Jansen's words:

It was in this state (under the influence of ketamine) that I experienced 'myself' as melded and intertwined with hundreds of billions of other beings in a thin sheet of consciousness that was distributed around the galaxy...Thus transpersonal events may be possible within the new physics, if subatomic events are involved in consciousness. Ketamine may be a drug which re-tunes the brain to allow awareness to enter the quantum sea. If this is indeed the case, then we may have to regard some of the reports of eternity, infinity, multiple universes and linkage with other beings as phenomena demanding a more sophisticated explanation than a brief dismissal as 'hallucinations and mental illness' requiring no further consideration (pp. 19-21).

Now the skeptical materialist might argue that Jansen has made a conversion from materialism to spiritualism because of his own psychedelic experience which has altered his brain chemistry and re-wired some of his brain circuits. In essence, they might contend that his psychedelic experience was so intense that he came to believe in its reality rather than accepting it as an hallucination or illusion. Furthermore, Jansen is basing his theory of the ketamine-quantum mind connection on the more exotic Copenhagen interpretation of quantum physics which, to his credit, is the dominant theory in the quantum field. However, EPR and Bohmian mechanics removes most of the mysticism from quantum mechanics, restoring causality and classical, logical understanding of the quantum world. Nevertheless, EPR and Bohmian theory have their short-comings as well, indicating that quantum theory is, as Einstein said, an incomplete picture of the reality of the unseen world of the very small.

Consciousness: Another Well-Kept Secret

The foregoing discussion on parapsychology and NDEs is indicative of the role of consciousness in physics theory. We have already discussed the role of consciousness of the observer in the outcome of quantum experiments and the role of consciousness in the anthropic principle derived from quantum theory. Defining consciousness is like defining time. We all know what it is until someone asks us to define it. It is something that is non-material in that it is not in itself tangible or visible. As Descartes said we are conscious therefore we know we exist. We have looked at whether consciousness can exist independently of the brain as some of those who have studied NDEs and psychedelics believe. The real mystery is how a non-material thing such as consciousness can be created by a physical organ – the brain, or, as some would say, how the brain channels consciousness of the mind or spirit. Some contend that machines will eventually come to consciousness, but since we don't know what causes consciousness, it is a false inference to say that because machines with artificial intelligence can perform many operations that the brain can, that those machines can become conscious and develop a personality. Much sci-fi has been based on machines, coming to consciousness and turning on their makers in Frankenstein fashion – including Hal the Computer in 2001 A Space Odyssey and Sky Net, the defense network in the Terminator movies, which threatens to extinguish human life.

Science fiction aside, it is not surprising that quantum physicists have taken the lead in the study of consciousness since quantum theory indicates that the observer's consciousness plays a determining role in shaping reality. Not surprising also is that physicist David Bohm, the mystic, was one of the first to wade into these murky waters claiming that quantum field theory, involving the implicate order, was the way forward in understanding consciousness. We have already discussed Bohm's belief in animism in which consciousness is an inherent property of all matter (inanimate and animate). Bohm, Penrose, Pribram and others have proposed a holographic or holonic model of the brain as a mechanism for the production of consciousness (Talbot 1991), but none seem to nail this jello to the wall. Quantum consciousness, as some have called it, involves everything from entangled particles (which can communicate at a distance), the mind collapsing wave functions, and microtubules in the neuron which are small enough for quantum phenomena to occur. Hameroff provided an hypothesis that microtubules found in the neuron would be suitable hosts for quantum behavior (Penrose 1995). Microtubules are composed of tubulin protein, and Hameroff proposed that the electrons in these tubules are close enough to become entangled (Hameroff 2008).

Now to quantize consciousness is a Herculean task indeed. How does one begin to quantize or break into units the size of Planck's constant something that is non-material? How does one quantize thoughts or feelings? We sometimes speak of the "stream of consciousness" which implies that consciousness is a continuum, rather than being composed of discrete, quantum units. However, quantum physicists have attempted to quantize time and space (two non-material entities), so it should be expected that there would be attempts to quantize consciousness also. I suppose it is possible to quantize some particles in the brain since the brain works on electrical processes. However, the impulse of a neuron, or brain cell, is not based on electron particles and waves as in electronics devices such as computers. The neural impulse is based on ions the size of atoms. The body takes salts (NaCL and KCL) and breaks them into positive and negative ions (Na^+, K^+, and CL^-). Nevertheless, one could contend that ions are subject to quantum principles. The positive ions are pumped to the outside of the axon membrane while the negative ions remain in the axon tube thus polarizing the neuron. When an excitatory neurotransmitter hits the neuron, the pores of the membrane are opened and the nerve impulse travels to the synapse where another excitatory or inhibitory neurotransmitter is released into the synaptic cleft thus stimulating the next neuron. I suppose one could say that when there is an excitatory neurotransmitter released, there is a quantum statistical probability as to whether the pores of the axon tube will open and allow the positive ions to pass to the inside in the same manner that particles pass through the double slit in quantum experiments. The ions could be said to be in a state of superposition, and the neurotransmitter collapses the wave making the ions actualize their potentials and assume a definite position. So, we could say that a neural impulse conforms to quantum probability and that the averaging of these random neural events adds up to fairly predictable, organized neuronal and brain function.

But neural transmission does not necessarily produce consciousness. Think of all the neural processes in the autonomic nervous system that take place without our consciousness such as the beating of the heart and the action of the hypothalamus in regulating glandular function by controlling the pituitary gland. Furthermore, Freud taught us that much of our brain activity is unconscious and that our consciousness is just the tip of the iceberg of deeper mental activity. Intuition occurs when the unconscious mind has found a solution to a problem and pushes it into consciousness creating those "eureka" moments. So, again, how do you get consciousness out of the brain and neurons? It is horribly complex, but I suppose we should keep trying.

Quantum consciousness and consciousness of the quantum world

Some scientists believe that the action of ions across cell walls involves particles too large to allow quantum action; therefore, microtubules are said to be small enough to be associated with quantum consciousness. Penrose argued that wave function collapse is the way to understanding consciousness; however, this process is subject to the uncertainty principle and unpredictability. To hedge on this randomness, Penrose 1999) developed a new type of wave function collapse that happens in isolation, and he named it objective reduction. He hypothesized that every quantum superposition has its own slice of spacetime curvature and that when these become separated by more than one Planck length they become unstable and collapse.

Now, a familiar mantra in physics is that the senses and mind have evolved to comprehend and interact with the macro world, rather than the quantum world. Thus, a Neanderthal throwing a spear at an animal and hitting his target is based on some innate, intuitive calculations in the brain as to how much to lead the animal based on its speed and the speed of the spear or arrow. As noted, some quantum physicists maintain that the senses and the brain are not adapted to understand and interact with quantum phenomena which operates on different laws and principles from the macro world. Hence the logic and rationality (some use the misnomer "common sense") that have enabled our minds to successfully interact with the macro world is not capable of understanding quantum weirdness. However, this line of reasoning does not take into account, the very different cognition that takes place in altered states of consciousness produced by meditation, hypnosis, or mind-altering substances. Thus, visions; dreams, mysticism, religion, near death experiences, and other altered states produce perceptions very much akin to quantum ideas as Capra, Bohm and others have recognized. Here are a few perceptual similarities between quantum mysticism and altered states of consciousness: 1) time travel, 2) relativity of time (one day is as a thousand years and a thousand years as a day), 3) soul travel indicates that the spirit, like a quantum particle, is not limited by time and space and can simultaneously be omnipresent in space, 4) ESP indicates that the mind, like entangled particles, can communicate instantly over long distances, 5) block time: there is a timeless realm where time doesn't flow but exists in an eternal present, and 6) extra dimensions: as in string theory, mysticism posits that there are dimension where natural laws do not apply. Hence, few, if any, of the ideas of quantum weirdness are without counterpart in altered states of consciousness that arose in human evolution long before quantum theory was developed. However, the scientific revolution has caused many modern humans to discard supernormal beliefs, but quantum physics seems to have caused a revival of the notion of another realm where the laws of nature we are accustomed to no longer apply.

Holographic Universe, Consciousness and Information Theory

Arguably, the most mystical of physics theory stemming from quantum theory is the idea that the material world is illusion or *maya* as Hindus call it. If the material world that we perceive through our senses (and instruments we use to extend our senses) is unreal and illusory, then "empiricism" and the scientific method is dead. If everything we see as matter and hard reality is a mere projection of mind and spirit, then Galileo may well be wrong about a 10-pound weight falling at the same rate as a one-pound weight, and Zeno is probably right. Zeno's logic told him that swift Achilles could never catch a turtle because every time Achilles moved to where the turtle was, the turtle had moved on. To observe Achilles catching a turtle, therefore, is an illusion, and the logical mind that tells you he doesn't catch the turtle is correct because the mind or spirit is the source of truth and not the physical world one observes. It is interesting to hear so many physicists arguing that the subject of their study, the physical world, is unreal and that there is no solid physical reality.

Here's what Richard Conn Henry, Professor of Physics and Astronomy at Johns Hopkins University, had to say about the relationship of mind and matter:

*Get over it, and accept the inarguable conclusion. The universe is immaterial-mental and spiritual...The only reality is mind and observations, but observations are not of things. To see the Universe as it really is, we must abandon our tendency to conceptualize observations as things (*2005*).*

Some of the less mystical physicists believe that hard matter is really a kind of compressed energy that is bound together by electromagnetism and strong nuclear forces. But, it would be just as easy to say that energy is high-speed, fine matter that can be slowed down to form larger chunks of hard matter. Either way it is said, it comes back to the idea that energy and matter are interchangeable as Einstein said. The presence of energy does not mean the absence of matter and vice versa – matter and energy are not mutually exclusive phenomena. The idea that energy is non-material and purely mental is the ultimate mysticism. What does Dr. Henry think he, as an astronomer, sees when he looks out upon the universe through a telescope? He might as well look within his own mind if indeed there is no real hard matter out there and everything is mental. It seems that mental telepathy would be a better way to study the universe, rather than observation. Like Henry, come quantum physicists indicate that when you search for the smallest unit of matter, there is no there, there.

"I refute it thus!" Samuel Johnson famously refuted Bishop George Berkeley's argument for the unreality of the material world by kicking a large stone causing a definite sensation of pain (Boswell 1823). Johnson's refutation might be considered simple, but it carries considerable weight. Dr. Henry, like Berkeley and others of such mystical bent, may intellectually doubt that hard matter is real, but I would wager that they conduct their daily lives as though matter is real. I would argue that we act on what we really believe. Assuming that Dr. Henry drives his car on a busy expressway, I would bet that he treats those multi-ton material objects (trucks) coming at him at high speeds as real, hard matter and that his senses do not deceive him about this reality. I would say he believes that, as Newton said, Force is equal to MASS times acceleration. To believe that these objects of the highway are not hard matter, but mere illusions, is to invite death and destruction. In evolutionary terms, our ancestors, who believed that material objects are real, survived and reproduced and thus gave rise to modern humans, most of whom believe that material things such as lions and tigers, rocks and trees are hard reality.

Some theorists have gone so far as to suggest that the universe we see out there is a holographic projection similar to the Holodeck in Star Trek, and Yoda's ethereal image being projected over great distances by the "force" in Star Wars. Dr. Fred Alan Wolf says: "The concept of the universe as a giant hologram containing both matter and consciousness in a single field will, I am sure, excite anyone who has asked the question, 'What is reality?' (Talbot 1992: back cover)." At least Dr. Wolf recognizes the interaction of *matter* with consciousness which is a reasonable concept.

Information Theory and Physics

Related to the idea that the universe is non-material is the notion that the universe is ultimately "information". But information is meaningless without something that it refers to. The first question that arises when someone says they have information is "Information about what?" Let's take a familiar piece of information attributed to Einstein: $E=MC^2$. That is information about the relationship between energy, mass and the speed of light. Without mass, energy and light occupying the physical world, $E=MC^2$ is meaningless. Or take another piece of information, H_2O. Without two atoms of hydrogen and one atom of oxygen existing in the material world, that formula has absolutely no meaning. The idea that the universe is just some kind of computer simulation made up of zeros and ones is _____ you fill in the blank. To manipulate information in a computer simulation, you have to have electricity, and to generate electricity you have to have some mass, such as magnets and wires made of copper or some other conductor. Neils Bohr is quoted as saying: "Everything we call real is made of things that cannot be regarded as real." In a recent Google Tech Talk, Ron Garret makes the case for the "Zero-Worlds" theory. He says that either there are infinite universes or there is no universe, and since the many-worlds theory is counterintuitive, then the zero-universe theory is more tenable. Both are equally untenable to yours truly. Garret goes on:

One classical universe is really untenable…We are our thoughts…We are a simulation running on a quantum computer…The universe I exist in is a good high quality simulation…What you perceive as physical reality is not actually real; it's actually an illusion (2011).

This is surely the height of mathematical metaphysics and could hardly be classified as science, in my view. Perhaps we are back to "I think, therefore I exist", but everything else is an illusion. Or perhaps we can question our own existence if there are no universes in which to inhabit.

As indicated above, since the coming of the information age, some physicists and other scientists have come to view the universe a computer simulation. While some use a computer simulation as a metaphoric model, others seem to take the model quite literally such as Ron Garret and Richard Henry. A more sensible view of information is taken by Paul Davies (Davies and Gregersen 2010) who says that information is one component of the matter, information and mathematics triangle, with information being the most important part. Davies speaks of top-down processing whereby one starts with information and proceeds to the concrete reality of matter-energy. Of course, he is talking about the inductive method which proceeds from general to specific. It is true that everything in the human realm is informational starting with sensory perception of the world, processing of that information using innate structures of the brain and cultural concepts. This processing of sensory information with organizational information might be called meta-information. The key difference between yours truly and the more mystical theorists, who do not recognize that information refers to a concrete entity, is that this author accepts the fact that human information is about an outside physical, objective reality. Without this assumption of a physical reality to which our information refers, empiricism is dead and so is science which demands physical proof for all theories. Without acknowledging the physical world, physics has indeed become metaphysics and perhaps mysticism. Here is my model of the relationship between information and material reality.

Primary Information: sensory data or extension of senses with technology. There is inherent information in matter and nature, but our perception of it is epiphenomenal. Thus, we must make a distinction between natural information and meta-information which is a representation of natural information.
Secondary Information: information for processing the raw sensory data or data derived from technical sensing. This includes categorizing and finding linkages between categories as in cause and effect. This processing involves both qualitative language and quantitative language (mathematics) that attempts to discover the hidden links between observable phenomena.
Tertiary Information: feedback from acting on primary and secondary information as in testing theories by systematic observation and experimentation and developing technology.

Now let's try to figure out how top down processing of information would be possible given these levels of information. In discovering the concept of gravity, Newton first had to observe material objects and derive the concept of mass, observe apples and other things falling to the earth, the moon going around the earth, etc. This would have been primary information, starting at the bottom, not the top. Then in the central processing part of his brain, he would have made a link among all these similar sensory phenomena to come up with the idea that there is a mutual attraction among masses. It is difficult to see how he would have started with the concept of gravity in top-down fashion, which would then have attracted his attention to falling apples and planetary motion.

The crux of the matter is that unless there is a real physical reality outside the human mind and information systems, there is no tertiary information - no answer from the physical world when we do an experiment. Then the mind is free to imagine string theories, quantum loop gravity, wormholes, etc. It is, to paraphrase Feynman, imagination without a straitjacket. When Ron Garrett uses polarizing filters to demonstrate constructive and destructive interference, he is getting an answer from the physical world of photons and light waves – *otherwise a thought experiment would have done just as well.* Information is innate in matter, but the way we represent that inherent information with words and math is an epiphenomenon and an imperfect representation of the natural information in matter-energy.

~Plato and the Greeks assumed that the material world is an imperfect representation of the perfect forms (concepts and theories) of the mind.
~Empirical science assumes that the theories produced by the mind are imperfect representations of the material world.

It is obvious to me that many physicists have returned to the Greek way of thinking based in pure rationalism. Of course, to deal with all these issues related to quantum consciousness and information theory goes beyond the scope of this book, but I would like to tackle these issues in later work. Suffice it to say that the study of consciousness has entered the realm of modern physics and has entangled physics in an interdisciplinary web where physicists cannot claim a monopoly.

Field Theory, Information and Consciousnes

As indicated previously, modern physics has become more and more abstract and mathematical and less and less about the physical world - some physicists even deny the reality of the physical world. Field theory is an excellent example of this trend. In Classical Physics, a field was conceived as a region of space in which material bodies exerted a force on each other – in a sense a field was a medium – some called it ether. Thus, Newton conceived of the space between celestial bodies as a gravity force field in which such bodies attracted each other. However, in Modern Physics, a field is not conceived as an array of particles or energy and, as such, is not physical reality. A field as now conceived is pure information or, as I have called it, disembodied information. Thus, it is a set of mathematical possibilities and probabilities which are described as virtual reality. Virtual reality is the probability that matter or energy will come into being or be manifested in a particular space.

In relativity, a field is comprised of spacetime. Space is defined as a vacuum or nothingness and time is a concept abstracted from change and motion. Although relativity denies that gravity is a force, it posits that massive bodies exert a force on spacetime which results in a warpage which in turn exerts a force on other bodies causing them to move within the grooves of spacetime.

In quantum theory, an orbital surrounding an atom is not a physical reality, but a virtual reality – a probability or potential that an electron will be present in that space. However, I would argue that an orbital represents a force field that attracts a free electron. Furthermore, in quantum theory, a wave is not a physical wave of energy, but a probability distribution that a particle could be present in a given location. This notion that a wave is a probability distribution is what gave rise to the otherworldly "Many Worlds Theory." In this theory, each statistical possibility in the wave distribution becomes a reality as new universes are created to actualize each possible reality.

In the Copenhagen interpretation of quantum physics, consciousness enters the picture of wave probabilities. The observer's consciousness is what collapses the wave of possibilities and makes only one of them a physical reality. When the experimenter makes an observation according to Schrodinger's parody, a cat, whose life has been hanging in the balances due to a 50-50 quantum probability, has its fate sealed by the observer's consciousness when the observer opens the cage and sees which state the poor cat is in – whether "wanted dead or alive".

Morphic Resonance and field theory

Rupert Sheldrake (2009) has written a book entitled *Morphic Resonance* in which he applies the concepts of quantum field theory to chemical and biological phenomena. Since chemistry involves quantum phenomena in the exchange and sharing of electrons in creating compounds, it is reasonable to deduce that such interactions of atoms involve quantum principles. Additionally, biological phenomena involve complex organic compounds such as proteins and nucleic acids which in turn include quantum interactions involving electro-magnetism. Not only do atomic and subatomic particles manifest quantum field properties, but compounds and organisms also manifest field properties at their respective levels, and the collective field properties of organisms, past and present, may determine the form a developing organism takes. For example, chemists around the world find that the melting points and crystallization points of chemicals change in tandem with each other even though physically separated. Of course, one must consider the possibility that these chemicals have come from the same sources and have had contact with each other. This non-material morphic field, according to Sheldrake is more important than DNA in determining development. Thus, his theory is more holistic than reductionist with top-down processing being more important than bottom-up processing. In other words, the whole is greater than the sum of the parts. A cell is more than its DNA, tissue is more than a collection of cells, an organ is more than a collection of tissues, etc.

Morphic Resonance is an abstract concept, but essentially it means that the universe has a memory and that these memories determine the form and function of phenomena at every level. This memory is seen as non-material information that exists in a field similar to the quantum field. Morphic resonance might be seen as the universe's data stored in "the cosmic cloud". For example, Sheldrake believes that DNA does not encode all the memory of a species which determines the development of the individual but that the species and ancestral memory is carried in a morphic field. Thus, he believes in the Lamarkian idea of transmission of acquired characteristics since an acquired characteristic would be part of the memory of an individual and therefore passed to the species field. I would argue that if acquired characteristics affect the sperm or egg, then these characteristics will be transmitted. If the acquired trait affects only the phenotype, and not the genotype, then the trait will not be transmitted to succeeding generations. Nevertheless, Sheldrake argues that "if rats learn something in one place, say a new trick in running a maze, then rats everywhere else should be able to learn the trick faster (p. 253)." This idea certainly presupposes a psychic connection among rats that is not bounded by space or time.

Sheldrake also believes that the brain does not contain the memories of the individual and that memories, again, are stored in the morphic field and that they persist even after the person's passing when the brain is dead. Hence his ideas are seen as supportive of non-materialist philosophies such as New Ageism which embraces the idea that the spirit is separate from the body and survives bodily death. His ideas have generally not been accepted in Western science but have gained traction in India because, perhaps, they are compatible with Hindu mystical philosophy. His idea of morphic fields of information is similar to the Hindu ideas of Brahmin, the oversoul, mystical oneness and reincarnation where memories are carried through several lifetimes. His ideas are also similar to Carl Jung's belief in the collective unconscious and archetypes; however, Jung believed that these species memories are carried through physical heredity.

Sheldrake also sees a close connection between the Morphic field and the quantum foam which makes up the field of space. In physics theory, particles and anti-particles are constantly bubbling up out of the vacuum and annihilating each other producing radiation. The information for this creative process must be embedded in the vacuum of space. As Sheldrake puts it:

Another possible point of connection between morphic fields and modern physics is through the quantum vacuum field. According to standard quantum theory, all electrical and magnetic forces are mediated by virtual photons that appear from the quantum vacuum field and then disappear into it again (2009:Kindle Edition Location 418).

Furthermore, Sheldrake perceives a strong connection between String Theory and Morphic fields. He claims that the extra dimensions of string theory could be where the information is stored that determine the forms we see in the 3-D world.

The following are some principles of quantum theory that Sheldrake has applied to the chemical and biological realms.

1) Like quantum field theory, Sheldrake posits that morphic fields are non-material entities of mathematical probabilities that determine the form and function of organisms as they use matter and energy in the processes of life. Hence, morphic fields are information fields; however, Sheldrake rejects information theory based on computer technology as too mechanistic to describe biological reality. He says that computer systems are closed systems, whereas organismic systems are open systems.
2) As in quantum physics, morphic resonance is like wave interference in which waves of equal frequency and in phase with each other amplify each other (resonate), and waves out of phase impede each other and may cancel out. Hence waves of species memories that reinforce each other by constructive interference are more likely to be manifest in the individual's development.
3) Since the morphic field is non-local in the sense of being omnipresent in the universe, then its information can be communicated instantaneously at a distance just as entangled particles can communicate instantaneously at a distance. This instantaneous communication refers not only to atomic and subatomic particles but to macro objects such as whole organisms in Sheldrake's view.
5) Just as the morphic field is not limited by space, it is also not limited by time. In the same way that some quantum physicists believe that information waves can come from the future, Sheldrake believes that information from the morphic field can come from the future experience of organisms and chemical interactions. This idea that morphic fields can be carried through time and space gives rise to his belief in parapsychology in which organisms, such as humans, can perceive events in the future (precognition) and events at a distance (clairvoyance). Like David Bohm, he also believes in psycho-kinesis (moving objects with the mind alone).
6) Sheldrake collaborated with David Bohm who felt that the morphic field is like Bohm's implicate order in which all things in the universe are interconnected. To Bohm, the explicate order is the phenomena that we experience in the empirical world which is fragmented and separated into categories. On the other hand, the implicate order is what lies beneath the surface and is the indivisible wholeness that is the ultimate reality.
7) Sheldrake's theories also mirror quantum indeterminacy and deny the existence of eternal laws which are so much a part of classical physics. He argues that if, at the moment of the Big Bang, there was no matter, energy, space or time, there could have been no eternal laws. Here he seems to contradict his concept that information does not depend on material media but is non-corporeal. If this is the case, then one could argue that the information on how the Big Bang was to unfold

was already present in the void. Sheldrake would probably respond that the universe made up temporary laws as matter and energy began to interact in space and time as the Big Bang unfolded and that these provisional laws are subject to change – they are not eternal. If this is his argument, then he contradicts the idea that the laws of nature are transcendent, existing in a morphic field rather than residing in matter. If the universe made up the laws as it evolved, then it would seem that these laws are immanent or intrinsic in matter-energy itself. In the materialist-idealist debate, Sheldrake seems to cross himself, vacillating from one to the other.
8) Sheldrake certainly speaks of morphic resonance as an eternal law of the universe. The argument hinges on how general or specific the laws of nature are conceived to be. Certainly, morphic resonance would evolve and change in specific ways, but the general principle of information being preserved in a non-material field is conceived as eternal by Sheldrake.

The whole idealist-materialist dialectic is a chicken-and-egg question – which came first consciousness or matter. Idealists might say that consciousness preceded matter and created it along with the laws that govern it. The materialist would probably say that matter came first and the laws that govern its interactions evolved into consciousness. The idealist-materialist debate can be seen in the controversy regarding acquired characteristics and the mechanism of heredity. The materialist would say that heredity is carried in the genes and thus characteristics that are acquired by an organism in its phenotype will not be transmitted to the next generation since the characteristic has not affected the genotype (the sex cells). The idealist, espousing the morphic resonance concept, would say that a characteristic acquired by an individual would become a part of the morphic field and the species memory and could be transmitted to future generations, regardless of whether the trait affects the phenotype or the genotype. Although Sheldrake presents evidence for this kind of transmissions of acquired characteristics in simple organisms such as fruit flies, the evidence for the human species seems to contradict this notion. The practice of circumcision among Jews has been going on for several millennia, yet each generation of Jewish males appears to have the foreskin intact despite this abbreviation of male genitals in a long line of their ancestors.

The irony of this non-materialist, non-mechanistic philosophy of morphic resonance is that when the scientific method is applied to it to prove it, there is the same demand for physical, material proof as in materialist theories. If there is a force exerted by the morphic field, science demands that it be manifested in some material form, such as loss of foreskin of succeeding generations of Jewish males.

It is also an interesting irony that Sheldrake's non-materialist philosophy of science has come under such intense fire by the materialists who have contradictory attitudes regarding morphic resonance and quantum mechanics. As indicated above, the morphic resonance theory shows a strong connection to quantum theory, yet quantum theory is generally seen as part of the materialist philosophy of science. It appears to this author that quantum theory partakes strongly of non-materialist philosophy. Such things as the consciousness of the observer determining the outcome of experiments, instantaneous action at a distance without any known physical medium, waves as mathematical abstractions rather than oscillating energy, Bohm's and Capra's quantum mysticism etc. are hardly forms of materialism. Furthermore, one photon going through the slits displaying interference with itself might be interpreted as the photon getting information from a morphic field. The quantum eraser experiment lends itself to morphic interpretation as does the idea of a pilot wave that guides a particle along a particular path. A dialogue between Sheldrake and Bohn bring out many of the corresponding ideas between quantum mechanics and the morphic field.

This contradictory attitude is shown by Dawkins, a rabid atheist biologist. Dawkins supports Lawrence Krauss's application of quantum theory to cosmology in which the universe sprang from nothing out of a quantum field, but he decries Sheldrake's morphic field as pseudoscience. However, quantum theory has *respectability* for sociological reasons rather than scientific reasons because it is accepted by mainstream physics. While I am not a fan of Morphic Field theory, I am pointing out the hypocrisy of physicists who espouse quantum mechanics which strongly suggests a morphic field and their rejection of morphic theory by Sheldrake.

Consider these analogies for understanding the relationship between Morphic Resonance and Idealism/Materialism.

Idealism is to Materialism as
Information transcendence is to Information Immanence*
Consciousness>Matter is to Matter>Consciousness
Organismic is to Mechanical
Holism is to Reductionism
Morphic field is to DNA
* Transcendence suggests that information is separate from matter and immanence suggests that information is embedded in the matter and not separate from it.

> One support for morphic resonance is seen in quantum, double-slit experiments. When many particles are shot through the slits over time, the dots on the screen arrange themselves in wave-like patterns of heavy concentration of dots (positive interference or resonance) alternating with light concentration of dots (negative interference). This fact suggests that the earlier particles were creating a morphic wave field which the later particles followed in their path to the screen.

Can materialist physicists rescue quantum weirdness from mysticism and spiritualism?

Cox and Forshaw (2011), are two young physicists who are staunch defenders of the faith. They protest that Modern Physics is a materialist science, not a spiritualist enterprise, despite evidence to the contrary. They emphatically declare that any mystical connection between quantum theory and mysticism is spurious.

Extrasensory perception, mystical healing, vibrating bracelets to protect us from radiation and who-knows-what-else are regularly smuggled into the pantheon of the possible under the cover of the word 'quantum'. This is nonsense born from a lack of clarity of thought, wishful thinking, genuine or mischievous misunderstanding, or some unfortunate combination of all of the above (Kindle Book: Location 83-86).

Yet they admit the following mystical claims of quantum theory.

Quantum theory does, admittedly, have something of a reputation for weirdness…Cats can be both alive and dead; particles can be in two places at once; Heisenberg says everything is uncertain (Kindle Book: Location 78).

What Cox and Forshaw leave out of their admission of quantum weirdness is that, in the Copenhagen interpretation, the observation (consciousness) by the experimenter determines the outcome of an experiment and determines whether the particle is here or there and whether the cat is dead or alive. It is difficult to demystify this claim and separate it from parapsychologists' claim that the mind can influence material reality in what has been called "psychokinesis." As indicated above, some physicists embrace the mystical and spiritual implications of quantum physics with great enthusiasm. In this author's estimation, those physicists who admit the inescapable connection between modern physics and the paranormal are much more honest and much less hypocritical. Cox and Forshaw seem to forget that the founding fathers of quantum physics were followers of Eastern mysticism, and when Schrodinger became disillusioned with the theory, he created a parody of it in his thought experiment of the cat in the box of quantum probability. Many physicists seem to take Schrodinger's parody seriously, but Schrodinger in the end declared this about quantum physics: "I don't like it and I'm sorry I ever had anything to do with it" (Gribbin 1984: preface).

Moreover, Brian Josephson, a physicist who invented the Josephson Junction, which represents a breakthrough in superconductivity and a significant step toward quantum computing, begs to differ with Cox and Forshaw regarding the physics-paranormal connection. In the late 70s, Josephson developed an interest in Eastern philosophy, meditation and higher states of consciousness, and, in the 1988, he set up long-running Mind-Matter Unification Project. Since that time, he has written on such topics as telepathy and homeopathy (Clegg 2014: p. 129). Josephson's interpretation of quantum mechanics as a path to understanding consciousness is so reminiscent of David Bohm's that I am tempted to call him the new "Bohm". Josephson's approach, which is in sharp contrast to Cox-Forshaw's approach, underscores the division in the physics community between those who admit the obvious connection between mysticism and quantum theory and those who vehemently deny it. To deny the connection is an attempt to shore up the image of Modern Physics as real, hard science and avoid the appearance of pseudo-science voodoo. But, let the chips fall where they may and call a spade a spade. If quantum theory leads to a theory of quantum consciousness, then so be it.

What this chapter has shown is that a study of consciousness is an interdisciplinary field including philosophy itself. However, we shall see in the following chapter that Stephen Hawking thinks that philosophy is dead and has no relevance in the modern, high-tech world of science. Despite the pronouncements of Hawking, this chapter and the next demonstrates that science comes down to philosophy. Science is, in itself, a particular philosophical system.

CHAPTER 3: PHILOSOPHY AND MODERN PHYSICS

> *Scientific theory begins in philosophy and ends in philosophy,*
> *and philosophy informs and guides every step in between.*
>
> *Physics is all about philosophy because the essential quest of physics is epistemological, that is the search for reality. Relativity and quantum physics seriously challenge our concept of classical reality and the classical scientific method.*
>
> *"The philosophy of science is as useful to scientists as ornithology is to birds."*
> *Richard Feynman (Trubody 2017)*

Richard Feynman, the quintessential scientist, fell out with philosophy after studying it for a while because of its seemingly endless quibbling over language and semantics which he considered to be trivial. Not to take Feynman's bird analogy (above) too seriously, but ornithology can be potentially useful to birds since bird-scientists are finding ways to protect our fine-feathered friends from the hazards of modern technology. Ornithologists have discovered that birds are being fatally drawn to wind turbines. As part of research conducted by Auburn University, bird-scientists are trying to develop radar and visual systems to help stop various avian species from striking wind turbines (Bryce 2016). In the same way, philosophers of science can help protect scientists from violating the scientific method by being more mindful of their assumptions and language. The scientific method is based on a philosophy of science that has been fine-tuned since the scientific revolution credited to Galileo. Furthermore, Feynman and other scientists, whether they are aware of it or not, are guided by some philosophy when they do science. It is better to be aware of those philosophical assumptions that guide one's work so that assumptions and premises can be evaluated in the light of science. It is the thesis of this book that Modern Physics has strayed from this time-honored method that has contributed so much to our knowledge and progress. We find that Stephen Hawking, like Feynman, holds a rather dim view of philosophy.

Is the Connection between Physics and Philosophy Dead?

As the foregoing discussion indicates, physics is very much connected to philosophy, especially epistemology and the philosophy of science. However, Stephen Hawking does not seem to recognize this intimate connection between physics and philosophy. In his 2010 book entitled *The Grand Design: New Answers to the Ultimate Questions of Life* (co-authored with Leonard Mlodinow), Hawking asserts in chapter one that "philosophy is dead".

Living in this vast world that is by turns kind and cruel, and gazing at the immense heavens above, people have always asked a multitude of questions: How can we understand the world in which we find ourselves? How does the universe behave? **What is the nature of reality?** *Where did all this come from? Did the universe need a creator? Traditionally these are questions for philosophy, but philosophy is dead. Philosophy has not kept up with modern developments in science, particularly physics. Scientists have become the bearers of the torch of discovery in our quest for knowledge. The purpose of this book is to give the answers that are suggested by recent discoveries and theoretical advances (2010: preface).*

Stephen Hawking has also asserted that fundamental questions about the nature of the universe could not be resolved without hard data such as that currently being derived from the Large Hadron Collider and space research. Although I admire Stephen Hawking for his fortitude and courage in not allowing a debilitating disease to imprison his brilliant mind, I think he went far beyond his element in these pronouncements, and perhaps he means that ***post-modern philosophy*** is dead, rather than the classical philosophy of science. Or, perhaps he was trying to be provocative to stimulate thought and discussion through these outrageous statements.

However, if Hawking literally means all of philosophy is dead, does he not know that physics itself is a branch of philosophy and that all branches of knowledge began with philosophy and is traceable to the Greek philosophers? Philosophy is the mother of all academic disciplines including the humanities and the sciences. This renunciation of philosophy is rather like a teenager saying that he doesn't need his mother anymore because her knowledge is old-fashioned and dead and should be buried with the past. Hawking's own discipline of physics was once called Natural Philosophy, and the "Ph.D." (Philosophical Doctor) in his title indicates that he is a "doctor of philosophy" in physics. Since the breaking of philosophy into various branches, including the sciences, general philosophy has remained as the discipline that examines the epistemological underpinnings of theory. This book is about that thought process, influenced by language, which is the foundation of all science. Hence it is a book about the philosophy behind science and particularly physics. Thus, it would be absurd to think that philosophers would have spent billions of dollars on particle colliders so that they could keep up with their offspring – physics.

This rather arrogant statement by Hawking is very much like Lord Kelvin's statement in his address at the British Association for the advancement of Science in 1900 in which he stated, "There is nothing new to be discovered in physics now. All that remains is more and more precise measurement (wikiquotes)." Hawking has indicated in other statements that physicists are now very close to a TOE (Theory of Everything) or grand unification of all four forces in nature. However, so far as I know, since the unification of electricity and magnetism, there has been only one other unification, namely, the electroweak, i.e., the uniting of the electro-magnetic force with the weak force that causes atomic nuclei to disintegrate.

Furthermore, Hawking's idea that all of physics is based on "hard data" rather than philosophical speculation is simply inaccurate. Much of physics is based on mathematical metaphysics, thought experiments and even mysticism as has been demonstrated previously. The Many-Worlds Theory, the Anthropic Principle, and String Theory could hardly be said to be based on "hard data". Even John Wheeler's Participatory Anthropic Principle is nothing less than mystical.

Moreover, to assert that physics can answer all the basic questions of human existence would suggest that not only do we not need philosophy, but we don't need any other academic discipline. This boast smacks of narrow discipline bias and tunnel vision. How do the findings regarding the nature of subatomic particles obtained from large colliders answer the questions of the meaning of life, the God question, or if there is indeed "a grand design" of the universe as the title of his book suggests? Hawking himself does not believe in a grand designer, so how can there be a design without a designer? One might speak of order coming out of the interaction of matter and energy resulting in the evolution of life, but order is different from design. Hawking has admitted that the origin of life on this planet was highly improbable suggesting that we really don't have a theory to account for it – so physics does not provide the answer to even this basic question - contrary to his pronouncement.

The "anthropic principle" is a weak attempt to answer the question of human origins, but is no better, in my opinion, that any such answer that has come out of general philosophy. Design implies a purpose, an end goal – in a word, teleology. Hawking should know that the most persistent questions of human existence, such as whether the universe was intended or came about by chance, go well beyond the scientific domain. Science can tell us *how* something works but cannot tell us *why* it works. The things that cannot be counted are sometimes the things that count most. Unless you are a follower of Pythagoras, you know that mathematics cannot unlock all the secrets of the universe. So this book is unashamedly about philosophy and more specifically, the philosophy of science and how we know what we know. Again, when physicists, like Hawking, attempt to answer the questions of the meaning of life through modern physics, they invite philosophers to enter the debate as to the legitimacy of the claims of physics. Anyone who has a Ph.D., which stands for "doctor of philosophy," can have a valuable say on philosophical questions that Hawking has raised. And, people without a Ph.D. who are serious students of life can also add valuable insights to these issues. Philosophy has not been totally supplanted by particle physics as Hawking arrogantly suggests.

My general critique in this book is that physicists have not paid nearly enough attention to the philosophical assumptions embedded in physics language and logic. It is interesting to note that Einstein, one of Hawking's idols, takes a point of view in direct opposition to that of Hawking regarding the value of philosophy in theory formulation. Einstein said:

I fully agree with the significance and educational value of methodology as well as history and **philosophy of science**. *So many people today—and even professional scientists—seem to me like somebody who has seen thousands of trees but has never seen a forest. A knowledge of the historic and philosophical background gives that kind of independence from prejudices of his generation from which most scientists are suffering. This independence created by philosophical insight is—in my opinion—the mark of distinction between a mere artisan or specialist and a real seeker after truth (Einstein. letter to Robert A. Thornton, 7 December 1944. EA 61-574).*

And regarding those scientists who made statements similar to Hawking's regarding the uselessness of general philosophy in the specialized field of physics, Einstein says:

How does it happen that a properly endowed natural scientist comes to concern himself with epistemology? Is there no more valuable work in his specialty? I hear many of my colleagues saying, and I sense it from many more, that they feel this way. I cannot share this sentiment. ... Concepts that have proven useful in ordering things easily achieve such an authority over us that we forget their earthly origins and accept them as unalterable givens. Thus they come to be stamped as 'necessities of thought,' 'a priori givens,' The path of scientific advance is often made impassable for a long time through such error (Einstein, 1916, "Memorial notice for Ernst Mach," Physikalische Zeitschrift 17: 101-02.).

Given Einstein's respect for philosophy as a necessity in formulating sound theories, now let's take a look at some of the philosophical questions that are foundational to physics. The key question with regard to theoretical physics is "What is reality?" Once, physicists decide on the answer to this question, then the next question becomes "How do we know what we think we know about this reality?" - in a word – epistemology. Epistemology then brings us face-to-face with the scientific method and how we obtain knowledge about nature. Is the research method which is used in physics the same method that is used in other sciences, or is the methodology of physics peculiar to its discipline? I have contended that physics uses a different methodology and logic which is not within the scientific domain.

It is interesting that Karl Popper, famous philosopher of science, agreed with the idea that some of modern physics is not scientific. For starters, he agreed with Einstein's objections to the Copenhagen interpretation of quantum physics but supported Albert Einstein's realist approach to scientific theories about the universe (Popper 1982). However, I would not consider Einstein's theories of relativity to be "realistic" since the abstractions of space and time are treated as real physical entities equivalent to matter and energy. Furthermore, there is much subjectivism in Einstein's relativity. Time is not objective and absolute but is personal and varies from one person to another depending on their speeds or the gravity field in which they are residing. And, the length of a rod is not an objective fact but depends on one's perception based on one's speed. Like Einstein, however, Popper strongly disagreed with Niels Bohr's *instrumentalism* which endorses the interpretation of scientific theories as practical instruments or tools for such purposes as the prediction of impending events (Popper 1962). Bohr's attitude was that since quantum theory works in terms of prediction, we should not question the aspects of the theory that are counterintuitive and illogical and that we should not seek a picture of the atom.

What is reality and how do we know it?

It would seem that the question of what is real in physics would be obvious. Physics is the study of the physical world; therefore, the physical world (matter and energy) is real and objective – it exists independently of our consciousness of it. However, there are physicists who subscribe to Eastern philosophy who believe that the physical world is *maya* (or illusion). It seems to me that the concept of what is real varies with the three major fields of physics: classical, relativity and quantum mechanics. Classical physics is a more common sense, intuitive branch and acknowledges the reality of what it aims to study – the physical world. Relativity acknowledges the reality of the physical world but ascribes equal reality to metaphysical concepts such as space and time. In classical physics, space and time are the background against which the interaction of matter and energy is played out. In relativity, space and time are not a passive background, they are actors in the drama of matter and energy actions and reaction. Thus, matter and energy can affect space and time (spacetime) and spacetime can affect matter and energy.

However, some physicists of quantum mechanics persuasion believe that matter is an illusion. Instead of hard matter, quantum theory sees the reality of the world as a set of probabilities and uncertainties – information, if you will. Hence, they seem to believe that abstractions (metaphysical things) are more real than hard matter. And those who are influenced by Eastern mysticism (Bohr, Heisenberg, Capra and Bohm) see reality in non-material and some would say spiritual terms. To some of them, reality is in the mathematical abstraction. A probability wave replaces a real wave, and information replaces hard reality. Neils Bohr is quoted as saying: "Everything we call real is made of things that cannot be regarded as real." Somehow what we perceive as macro matter is made up of things that are not materially real – this quantum reality is pure incorporeal, disembodied information. However, some believe that this quantum information adds up to or averages out to an illusion of macro, hard matter, but is, in reality, a holographic type projection. Garret says that "We are a simulation running on a quantum computer so that the universe is a good high-quality simulation. What you perceive as physical reality is not actually real; it's actually an illusion" (2011).

Despite these seeming denials of an objective, physical world, science demands physical proof of any theory that is proposed – a mathematical proof or thought experiment is not sufficient to support theory. Now if one does not believe in the reality of the physical world, how can there be scientific proof of any theory? Even parapsychology requires physical proof for psychic-spiritual phenomena. If one claims to be a psychic with clairvoyance, then one must perceive a physical fact without sensory intervention, and that extra sensory perception must be verified by checking to see if the physical fact exists as predicted. Hence a field of study which indicates that the physical world is not real is not a science – it is mysticism, religion or perhaps one of the humanities.

But then there are physics theories which indicate that one creates his own reality or that reality is relative to one's point of view. In relativity, what is true about speed, time and space depends on one's perspective (or frame of reference, if you will). In the participatory anthropic principle, not only does one's observation determine the results of an experiment, but one's observation creates the universe and its history going backward in time. This idea brings into play the role of consciousness in scientific theory and whether there is an objective reality outside our consciousness which we can know through our reason and senses. The scientific method is predicated on an objective, physical reality which existed before human consciousness and exists independently of consciousness. These issues are dealt with in greater depth in other parts of this book.

Scientific Realism

If I understand the essence of Scientific Realism, then my philosophy of science seems closely aligned with it because Scientific Realism posits that there is an objective, physical world that is independent of human consciousness. Without this assumption of an independent reality, science is not possible in the traditional definition of science.

According to scientific realism, an ideal scientific theory can be taken literally and must meet the following criteria:
1) The claims the theory makes are either true or false, depending on whether the entities talked about by the theory exist and are correctly described by the theory. This is the semantic commitment of scientific realism. This claim relates to the **falsifiability** of hypotheses and the importance of language in accurately describing the world.
2) The entities described by the scientific theory exist objectively and mind-independently. This is the metaphysical commitment of scientific realism. This is the ultimate basis of empiricism that theories (products of the mind) must conform to the physical world as systematically perceived by

the senses or instrumental extension of the senses.
3) There are reasons to believe, based on empirical evidence, that some significant portion of what the theory says is true. This is the epistemological commitment (Leplin 1984).

I would add a fourth criterion to scientific realism, related to the idea that objective, physical reality is independent of the mind and consciousness. **Language and math are approximations of the physical world** - they are not the physical world any more than a globe is the earth. Language and math are the only tools we have to describe the physical world and great care should be taken to ensure that they represent nature as accurately as possible. Max Tegmark is quoted as saying: *"All structures that exist mathematically exist also physically."* This statement indicates that there is no distinction between physics and metaphysics, the physical world and the mind. This is a very Pythagorean statement that nature is math and math is nature, and the underlying assumptions is that the map is the same as the territory.

Logical positivism (or rational-empiricism) was the original philosophy of science in the twentieth century and the forerunner of scientific realism, holding that a sharp distinction can be drawn between observational terms and theoretical terms, the latter capable of semantic analysis in observational and logical terms. Thus, Scientific Realism represents an attempt to fix some of the perceived flaws in logical positivism.

The essential feature of scientific realism which I find attractive is the belief in the reality of objective, physical world – a concept which is partially lost in relativity and seemingly completely lost in some aspects of quantum theory. It is the belief of this author, contrary to Hawking's belief, that philosophy should play an active role in evaluating physics theory. If logical positivism or rational empiricism is the basis of the scientific method, then physics theories should be systematically evaluated on these grounds. The philosopher does not have to understand the math or the technicalities of a theory to evaluate its philosophical underpinnings and assumptions. One does not have to understand the complex math of String Theory to evaluate whether the assumption of one-dimensional strings vibrating in 11 dimensions is rational or empirical and therefore scientifically realistic. If a theory, such as String Theory, hits glitches and reveals no supersymmetric particles as predicted, then the foundational assumptions of the theory should be evaluated.

Einstein's Philosophy of Science – Empiricism or Rationalism?

Imagination is more important than knowledge. Albert Einstein

Einstein's philosophy of science is a curious and contradictory mixture of rationalism (Greek rationalism to be more precise) and empiricism. On the one hand, he would say that there are things one can know through logic without observation, and other things that must be known through empiricism (observation) without logic. To illustrate this, consider that Einstein, by and large, was not an experimentalist - he was a theoretician. His favorite method of creating a theory was to engage in a grand thought experiment or *gedanken*. In the *gedanken*, his assumption was that the truth could be known by pure logic. For example, if one starts with the assumptions that the speed of light is constant and that the laws of physics are the same in all inertial frames, one can deduce that when something moves inertially (at a constant speed), time dilation and space contraction occurs which becomes apparent at high speeds that are near light speed. This type of thinking is similar to Greek rationalism which assumes that truth is in the mind and can be teased out by pure logic. Thus, Aristotle predicted that an object weighing 10 times more than another object will fall to the earth 10 times faster. Aristotle didn't think it was necessary to try this out in the real world because logic is superior to empiricism in Greek philosophy.

On the other hand, Einstein indicated his belief in empiricism – albeit limited and naïve empiricism. To begin with, he says that there is no observation that can inform a traveler on a ship, moving at a constant rate of speed, whether s/he is moving or not. The empirical fact (or sensation) that Einstein points to is that the passenger on a ship cannot "feel" the difference between inertia of rest and the inertia of motion; therefore s/he cannot know if s/he is moving or not. Thus, the passenger's feelings (or lack thereof) is the empirical basis of the experiment, and **s/he cannot use logic** to figure out whether s/he is in motion or not. For example, the passenger could use logic to look out the window at the ground to see that the ship is moving in reference to the earth frame or perform an experiment on the ship to detect movement or remember that the ship started by accelerating to a certain speed and then leveled off. The passenger must rely on feelings based in kinesthetic senses alone and must be blind-folded and unintelligent so that s/he does not figure out that the ship is moving. Of course, the blind-folded, unintelligent observer is necessary to show the discrepancy between the views of a ground-based observer in inertial rest and the observer in flight in inertial motion. Because the observer on the ship can't know s/he is moving, his/her clock reads differently on the ship than it does to the ground observer and thus since the light appears to travel faster on the ship in motion, this cannot be the case since the speed of light is assumed to be the same for all inertial observers. Thus, the time on the moving ship had to change in order to keep the speed of light constant. And if time slows down, space must contract to keep the speed of light invariant.

So we see that, for the most part, Einstein was a Greek rationalist rather than an empiricist. Now, of course, Einstein's assumption of the constancy of the speed of light was not hatched in a vacuum. This assumption was to some extent based on the equations of Maxwell and the work of Lorentz who developed the idea of length contraction for moving objects. The overall conclusion is that the moving ship cannot add to the speed of light even though it is carrying a light clock; however, the moving ship does cause time to slow down, but only in the view of the ground observer, so the motion of the ship does affect light – just not its speed. We shall speak more of relativity and Einstein in the book on Relativity.

Karl Popper and Scientific Realism

The philosopher of science who comes closest to my view of science is Karl Popper. Popper believed in scientific realism in which there is mind-independent reality or objective reality. Thus, he criticized the *subjective* Copenhagen interpretation of quantum physics which indicates that the observation of the experimenter determines the location of a particle whose location was *uncertain* before the observation (1982). This uncertainty principle with its mind-dependent outcome has provided justification for all manner of mystical theories of the paranormal including New Age ideology. While some theorists, such as Bohm and Capra, have embraced the implications of the "consciousness creates reality" principle of the Copenhagen group, others have staunchly defended quantum theory, denying that there is any connection between the uncertainty principle and parapsychological phenomena such as psychokinesis. These deniers seem, for lack of a better word, disingenuous, for denying this obvious connection between the quantum world and the mystical.

Bohr, in saying that one cannot separate observation from objective reality is implying that observed reality is the only reality we can know. For example, when an experimenter observes a wave, the wave always becomes a particle. Thus, the particle-wave duality is our reality – when we are not looking, the quantum is a wave; when we looking, it becomes a particle. What Bohr fails to realize is the distinction between active observation and passive observation, and since subatomic observation generally must be active and obtrusive, then there is interference with the phenomenon. Let's make this distinction a little clearer.

1. Active observation: affecting the object in the act of observing it, e.g., hitting a particle with light to determine its location can change its speed, location and trajectory.
2. Passive observation: not affecting object in the act of observing it. This can be achieved more easily in studies of macro things, e.g., using a hidden camera to record natural behavior of wildlife or human beings or time lapse photography of an eclipse.

This author also agrees with Popper's falsification principle, that is, for a theory to be scientific, it must be structured in such a way that one outcome of an experiment or observation will support the theory and another result will falsify it. Quite often, Popper points out, theories are designed in such a way that they are insulated from falsification. An excellent example of this is the Many Worlds Theory of quantum physics which is discussed in a later book. The theory denies that the quantum wave collapses as in the Copenhagen interpretation. Instead of collapsing several probabilities into one definite outcome, each possibility in the wave is realized by the universe splitting so that each probability is made real in some universe. The theory goes on to indicate that these other universes cannot be detected because of decoherence or the scattering of the waves. Popper would probably say this theory is insulated from falsification because it can never be tested since the predictions of the theory (other universes) can never be seen.

Popper also criticizes Bohr for *instrumentalism* which means that if a theory works, it does not need a rational principle to show why it works. Bohr's idea was that quantum mechanics yields accurate predictions and so there is no need to explain quantum weirdness such as the uncertainty principle and a particle being in several places at one time. His attitude seemed to be - If it works, it doesn't have to make sense to the rational mind. Popper, on the other hand, believed that a theory should not only yield accurate predictions but should not violate commons sense. An excellent example is that the Ptolemaic heliocentric theory with all its fudge factors and epicycles yielded more accurate results that the initial Copernican theory, but despite these accurate predictions, geocentrism did not conform to other observations connected by rational principles which showed heliocentrism to be the way the solar system works..

Since it is rare in science to find a theory that yields 100% accurate predictions, support for a theory becomes statistical. If an hypothesis can deviate from chance by a significant amount, then it is considered provisionally supported. For example, if an hypothesis yields a departure from chance at the .01 level, that means that there is a 1 in a hundred chance that the experimenter could get these results by chance alone. Einstein might say that "God doesn't play at dice" so the outcome would be 100% certainty if one knew all the variables and could measure them accurately. Thus, his philosophy of science was deterministic. On the other hand, quantum physicists of the Copenhagen interpretation would say that nature does play dice and that the level of uncertainty in not just a result of imperfect knowledge by the scientist, nature does involve a level of randomness and chance, especially at the quantum level, and this translates to uncertainty at the macro level of reality.

It is in the heat of this debate between Einstein (determinism) and Bohr (non-determinism) that Popper enters the picture with his *propensity theory of probability* principle. Popper's propensity theory holds that probabilities should be based on objective potentials rather than subjective estimates about probable outcomes. Popper probably saw that quantum physics involved subjective probabilities, rather than objective ones. Thus, rolling a balanced die (dice), one should get a six, $1/6^{th}$ of the time in a large number of trials since the die has six sides. However, in a single trial with a balanced die, one can state objectively what the possibilities are.

Falsification problem of induction: Popper is critical of the process of induction, i.e., generalizing from specific observations to general theories. As an example, just because the sun has risen every day in human experience does not mean that it will always rise. Of course, science tells us that the sun will continue to rise as long as the earth spins and the sun remains stable. However, we are told that the sun will eventually become a read giant and swallow up the earth, so that it will cease to rise for the earth at that time. The theory of the sun not rising one day might be based on induction from knowledge of the sun's dynamics. However, the **deductive method** of science also yields problems. If a scientist begins with an hypothesis and seeks data that will either falsify or support it, the human tendency is to filter the data so that non-supportive data are ignored or denied. This is what I call deductive bias. However, a cure for this problem is in double blind analysis of the data and peer review.

Thomas Kuhn's View of Science as Subjective and Social Constructionist

Standing in opposition to Popper's ideal objectivism in science is Thomas Kuhn, who believes that so-called science involves more of a political and sociological process than the ideal systematic scientific method. Once a paradigm (a set of integrated theories) is established, it becomes connected to the political establishment and resists change by mavericks and innovators because funding of research goes to those theories that enjoy widespread acceptance and respectability by professional scientists. To apply Kuhn's theory, we might compare the Copenhagen interpretation of quantum physics to parapsychology. Although the Copenhagen interpretation indicates that the observer's consciousness determines the outcome of an experiment, when parapsychologists claim this idea as evidence for psychokinesis, many in the physics establishment take offence at this connection to what they consider pseudoscience. Accordingly, Wheeler wanted to ban parapsychologist from the American Association of Science. Banning parapsychologists was not so much based on lesser evidence for the "psi" factor than the wave-collapsing observation of the experimenter as it was that physics has much "respectability" for its professionalism and parapsychology has little. Thus, Wheeler was acting on sociological, rather than scientific grounds. Furthermore, Halton Arp was prevented from using telescopes to find evidence for his theory of electromagnetism as the prime move in cosmic dynamics rather than gravity, so he moved to Germany to find a more tolerant attitude for his deviant theory.

Similarly, Kuhn says that the Copernican Revolution in its beginning did not offer more accurate predictions of celestial events, such as planetary positions, than the Ptolemaic system, but instead appealed to some practitioners based on a promise of better, simpler, solutions that might be developed at some point in the future. Kuhn's insistence that a paradigm shift is an admixture of sociology, enthusiasm and scientific promise, but not a systematic scientific method, caused an uproar in reaction to his work (2012).

Perhaps an Hegelian synthesis of Popper and Kuhn's ideas would come closer to the truth than either extreme. To begin with, I believe that Popper is saying that mind-independent, objective science should ideally be the basis of the scientific enterprise. However, while Popper believed that Einstein's relativity followed this process, he believed that the Copenhagen interpretation did not. Hence, he proposed an experiment to refute the Copenhagen philosophy. Kuhn, on the other hand, is saying that science is not in actuality conducted in the ideal Popper way, but is much influenced by politics, sociology and groupthink. In other words, scientific theories are not hatched in a vacuum and are not maintained in a vacuum – there is a socio-political context of which scientists are a part. Perhaps Kuhn would agree that for a paradigm to shift, the older generation holding the traditional view must die off to allow a younger set to establish their paradigm which they then defend against falsifiability with the same intensity as the older generation. If one looks at the current status of Supersymmetry and String Theory, one can see both theories of science in operation. The fact that both theories held sway for so long with no evidence to support them is testament to Kuhn's philosophy. The fact that no real evidence has been found for either theory, leading to a disenchantment with these theories causing some to abandon them in favor of new ideas scores one for Popper. In both these theories, sociology and the scientific method play a role in how science is done. When scientific theories are open to review, the lack of evidence for a theory will ultimately lead to its demise as we are seeing with String Theory and Supersymmetry. The scientific establishment will not exercise infinite patience for a theory that bears no fruit of evidence.

The Sokal Affair: A Test of Post-Modern Critique of Science as Social Reality

The Sokal hoax was pulled off by Alan Sokal, a physics professor at New York University and University College London. In 1996, Sokal submitted an article entitled "Transgressing the Boundaries: Towards a Transformative Hermeneutics of Quantum Gravity" to *Social Text*, an academic journal of postmodern cultural studies. In the article, he pretended to criticize the theory of quantum gravity as a social construct, giving nonsense information, citing individuals who had published works in the *Social Text* journal and playing to the leftist ideology of postmodernism. The article was accepted for publication, and later Sokal published a disclaimer saying that the article was a hoax and that much of the information was nonsense.

What Sokal proved was that postmodern academicians are certainly influenced by the cultural context in which they operate, and that bias determines what they will publish and accept as truth. Postmodern philosophers do indeed have their social constructs of reality. However, the hoax did **not** prove that scientists are not subject to the same cultural and political milieu in which they find themselves. Moreover, the hoax did not show that scientists are not biased in what they publish and accept as truth by their prevailing paradigms. Physical science does not live in a vacuum any more than the humanities and social sciences, and their objectivity is tainted by prevailing cultural ideas inside and outside science. What Sokal ignores is that some so-called valid theories of physics also deny mind-independent, objective reality. The Copenhagen interpretation of quantum physics clearly supports a reality that depends on the consciousness of the observer, and this theory is perpetrated in physics through a cultural and linguistic process as in the humanities.

Sokal in the *Lingua Franca* article (1996) in which he revealed that he had pulled a fast one on the gullible editors of *Social Text*, said that:

Anyone who believes that the laws of physics are mere social conventions is invited to try transgressing those conventions from the windows of my apartment. (I live on the 21st floor.).

Of course, what Sokal is referring to in this invitation to take a flying leap is Newton's law of gravity from Classical physics which **does** acknowledge an objective reality. However, Modern Physics is a different animal – and many of its theories are highly subjective. Thus, I would challenge Sokal, in his own apartment, to find the seven extra dimensions called for in String Theory and to perform Schrodinger's cat box experiment without resorting to the observer effect to determine if the cat is dead or alive, and perhaps to empirically determine if the cat is both dead and alive in different universes in the Many World's interpretation. We won't even mention the "Wigner's friend thought experiment" as adding to this subjective nonsense. And, if he wants to try jumping out of his own 21st floor window, perhaps he could employ the theory of negative gravity on the way down to save himself. These subjective theories of Modern Physics have been published in physics journals – not as hoaxes – but as real science. However, Sokal would probably contend, against all evidence to the contrary, that these theories are hard, objective science. Sokal needs to write a similar hoax of these subjective theories, using advanced mathematics and sophisticated-sounding physics jargon, to see if he can get them published in a physics journal – actually they already have been published as indicated before. It seems that Sokal's trick is a matter of the pot calling the kettle black. Science, like the humanities, is much influenced by the culture and politics within the field and in the wider society, but that does not mean that there is no objectivity in physics and other sciences. The truth, as usual, is somewhere in-between.

The Bogdanov Affair – a Reverse Sokal Hoax?

Perhaps Sokal should read about the Bogdanov Affair detailed in Peter Woit's book *Not Even Wrong (2006)*. Woit records an incident, which turned out not to be a hoax in the sense of being intentional deception. However, it did show that an article that a majority of physics regarded as nonsense got published in five physics journals, and many physicists couldn't tell whether it was a hoax or real physics. These physicists could not determine whether the article was science or pseudo-science until informed by leaders in the physics field who said it was nonsense. Sounds like social construction of reality to me. What the article underscored is a failure in the peer review process in physics journals to detect what turned out to be the kind of sophisticated-sounding, mumbo-jumbo that Sokal had perpetrated on the naïve reviewers of a post-modern philosophy journal. One of the Bogdanov brothers was flunked in his Ph.D. dissertation because of work that went far outside the parameters of String Theory. However, his Ph.D. advisor told him that if he could get his work accepted by in a peer-reviewed physics journal, he would pass the dissertation. The Bogdanov brothers proceeded to get three articles (that were almost identical) published in five different physics journals. Woit says of the articles: "…it became clear that it was a rather spectacular piece of nonsense, a great deal more so than anything I had previously seen in a physics journal (p. 215)." What makes it even hokier is that some physicists didn't recognize it as nonsense until it was labelled as such by the so-called experts. The following was reported to Woit by someone visiting the String group at Harvard.

So no one in the string group at Harvard can tell if these papers are real or fraudulent. This morning, told that they were frauds, everyone was laughing at how obvious it is. This afternoon, told they are real professors and that this is not a fraud, everyone here says, well, maybe it is real stuff (p. 214).

Does this confusion about whether the work of the Bogdanov brothers was rubbish or real science reveal anything about what Sokal had to say about post-modern philosophers who claim that much of science is "social construction" rather than objective science? The Harvard group couldn't tell whether the Bogdanov work was rubbish until they were told it was a hoax by some authorities and then changed their minds when they learned that the brothers were real professors of physics in France. As Woit points out, there is so much unscientific work published on String Theory that one cannot tell the difference between rubbish and real science – perhaps there is no difference when it comes to String Theory – and that accounts for the confusion.

It sounds as if physicists are subject to the same human tendencies toward group think and social construction of reality as are post-modern philosophers. It is a human tendency that everyone is susceptible to – including some physicists, like Sokal, who seem to believe that they transcend subjectivism. It would seem that this incident, contrary to Sokal, gives credence to the postmodernists' claim that scientists are, to a great extent, susceptible to groupthink and social construction of reality.

> Nutshell: Sokal got the first laugh from his prank, but the postmodernists should get the last laugh because of the Bogdanov Affair. This fiasco clearly supports the very claims the postmodernists were making – that much of science, parading as objectivity, is actually a social construction.

Common Sense, Empiricism and Modern Physics

Quite often, you hear it said that modern physics **transcends common sense**, or that physics theory is counterintuitive. Feyman said: "I can safely say that no one understands quantum mechanics." Others say that modern physics cannot be understood rationally and must be accepted on faith because the math says it is so. Still others have said that modern physics must be understood in terms of Eastern mysticism rather than Western rationality. Many times, when a theory leads to absurd conclusions, the escape hatch is that the theory is counterintuitive and cannot be comprehended with common sense. But common sense is such a vague term that it has little meaning especially in scientific parlance.

Just what is common sense? There are two meanings.

1) Common sense is sense we hold in common, that is, our common culture, particularly folk culture, which may embed superstition or fallacious thinking, e.g., the earth is the center of the universe – believe it or not, some people still believe in geocentrism.

2) Common sense can also mean *practical sense*, i.e., trying things out to see if they work and are therefore true. When Aristotle said that men and women have a different number of teeth, *common sense* would dictate that he should look into the mouths of several men and women and checked his theory out. Hence, the practicality of common sense is similar to the experiment (empiricism) in scientific theory, i.e., try your theory out and see if it works and is true. The basic difference is that the experiment involves controlling variables, so the experimenter can ideally tell which one is the cause; while common-sense practicality involves a trial usually without controlling extraneous variables. However, practical, common sense exercised by Galileo is the forerunner of the controlled experiment.

In modern physics, rejection of common sense also means that the sensible, understandable concepts of classical physics cannot be applied to the quantum level. For example:

The *uncertainly principle* is that you can't know the position and speed of a particle simultaneously. The more you know about the particle's location, the less you know about its velocity and vice versa.

Classical Physics based on common sense indicates the opposite about location and velocity, i.e., you can't know the speed of anything unless you know its position (or location) in at least two points and the time it took the object to travel between the two points. The more you know about an object's positions, the more you know about its velocity. This statement about macro-reality directly contradicts Heisenberg's statement about micro-reality. Which statement is more scientific, and which is more mystical? Recall that Heisenberg was a believer in Eastern mysticism.

Consider also that relativity and quantum physics not only contradict classical common sense, they contradict each other. Common sense says they cannot both be right, but in mystical quantum theory, contradictions can both be right, e.g., a cat can be dead and alive at the same time. Consider also that relativity says that nothing can go faster the light, but Quantum physics says that entangled particles can communicate with each other over great distances instantaneously (faster than light). Can both these theories be right – only if one can justifiably say that the two theories apply in different circumstances? However, at this point one must be careful of hedges and patches to theory similar to Ptolemy's epicycles, invented to make geocentrism work.

Determinism vs. Non-determinism

Physics is divided into three tribes each of which have different ideas as to whether the universe is determined or accidental, orderly or chaotic. The classical physics of Newton and Galileo is said to be deterministic in the sense of believing in cause and effect. If macro-phenomena are determined, then they are predictable if we knew all the variables that cause them to manifest as they do. Relativity is also said to be deterministic although the causes of things such as gravity are abstract, rather than physical. On the other hand, quantum physics, especially the Copenhagen Interpretation is decidedly non-deterministic. In other words, quantum phenomena are random and unpredictable, and because they are unpredictable, they are non-deterministic. The assumption is that if you can't predict a phenomenon, it does not have an order – if you can't know the cause, it doesn't exist.

I would contend that quantum phenomena are deterministic if we could know all the variables or causes, but we can't. It is not that these causes don't exist, it is our inability to see into the unseen world of the atom that prevents us from being able to know the causes and make predictions. Actually, things are not totally predictable in the macro-realm either - because we can't know all the variables and causes, and even if we know the causes, we can't always quantify them or know how they will combine to produce a specific effect. Take the weather, we probably know most of the causes of changing weather, but even with Doppler radar, we cannot predict exactly what the weather will be at a given time. However, we can predict a pattern and the probability that a certain weather event will occur. When a weatherperson says there is a 70% chance of rain, that is about as precise as s/he can be, and we would expect rain in 70 of 100 days in which we heard that prediction. Chaos theorists tell us of the "butterfly effect" in which the flapping of a butterfly's wings can be the deciding factor of which way the weather goes. For this to happen, forces must be in near equilibrium, and the butterfly's slight perturbation of the air causes a tornado which would not have occurred without the butterfly. These small factors are hard to plug into our computers since programs deal with the major factors. Furthermore, our minds are limited to 5-7 bits of information that can be held in our minds at a time; therefore, we can only take two or three variables into account at one time.

The weather is no different from quantum phenomena, it appears to me. Quantum theorist may not be able to predict the behavior of a single subatomic particle, but they can predict a pattern and probability of atomic behavior as in a double-slit experiment. A pattern suggests an order and a cause in quantum phenomena as well as macro phenomena.

Demystifying Uncertainty: Particles, People and Probability

God does not play dice (Einstein)...or does s/he?

Quantum physicists speak of **superposition** *which indicates that a particle can be in two or more places at once, or as Feynman said, a particle takes all possible paths and goes through both slits in the double slit experiment. Surely what physicists mean is that the particle has a potential or probability of being in a number of places and a potential for going through either slit. Hence, we should make a distinction between potentials/probabilities and actualities as we do with macro phenomena. Perhaps we should say there is a difference between potential energy (possibility) and kinetic energy (actuality).*

Does God (or nature) play dice or not? It seems to me that the QUANTUM world is much like the macro world in terms of indeterminacy and probability. The question for both the micro world and the macro world is whether the unpredictability and apparent chaos comes from our ignorance of all the variables or whether nature itself is random at certain levels. Quantum physicists claim that it is impossible to predict the behavior of individual particles because their behavior is random and chaotic – although the behavior of an aggregate of particles may average out to a discernible pattern. We in the social sciences do not find this quantum strangeness "strange". The same indeterminacy is true of people – they behave much like particles. We cannot predict the behavior of an individual with complete certainly even if we know the person's genetics and environmental influences. We can, however, predict a pattern of behavior if we know the person's history, and we assume that those patterns will continue into the future. The same is true of group behavior. We can predict patterns of group behavior but not for any particular point in time. For example, we can know that murder is more likely to be committed in certain areas of a city than others, but we cannot predict an exact time and place where the next murder is going to occur. However, we can say that murder is more likely to occur at certain times of the week (weekends) and certain seasons of the year (summer). And, we can say there is a kind of probability wave that builds in an individual and community. A wave of frustration and anger can be seen in communities where there is economic and social deprivation relative to other communities. And if there is a downturn in the economy, that wave may exhibit positive interference with others experiencing the same frustrations. When the wave builds to a critical point, the wave will collapse, and a murder will be committed – or perhaps a riot will occur. Someone may be scapegoated for the frustrations of another, or there may be conflict generated from competition for scarce resources.

Now, just as a particle is said to have no definite location or status until an observation is made of that particle, so it is with society. In a sense, a crime has not been committed until a person is arrested and convicted beyond reasonable doubt. In our system, a person is innocent until proven guilty; therefore, the accused has not committed a crime until the observers (the courts) make a definite decision that the accused person is indeed guilty of a crime and a label is placed upon him as a criminal. Again, we see that people, like particles, do not have a definite status until an observer or observers make it so.

Furthermore, twins could be said to be **entangled** genetically so that even if they are separated at birth and raised in very different environments, they show amazing similarities when brought back together in adulthood. They show a high correlation in IQ, personality traits and even in idiosyncratic gestures. If one twin is schizophrenic, the other one has a high probability of having the same mental disorder. So it is with twin particles that are separated and show similar traits at a distance.

Some people, including psychologists such as Jung, perceive non-local variables that influence human personality. Astrology, as an ancient belief system, posits that the constellations that appear to form the backdrop for the sun, moon and planets at the time of one's birth may influence an individual's core personality. Such non-local variables, however, might relate more to the season of the year that a fetus was carried in the womb and would thus be a local variable that has some influence on the personality.

God does play dice with biology and sociology as well as physics. Mutations in biology, which are said to be caused by quantum events (radiation disorganizing DNA) follow the uncertainty principle. It is impossible to know when and where mutations will occur and if they will be adaptive or maladaptive. And, we all know that the weather has random, unpredictable elements; however, a pattern can be predicted, and that pattern reveals itself as climate. Chaos theory applies to all of nature and there is randomness in all phenomena in nature which makes exact prediction impossible for the scientist and the prophet. Hence, there is nothing mystical about unpredictability in quantum phenomena - unpredictability is present in macro phenomena as well.

In sociology and physics, a theory is never complete [whether a Theory of Everything (TOE) or a theory predicting the length of the long toe] because no theory can take into account all the variables that affect a given phenomenon. As Feynman says - all our theories are but approximations. For example, General Relativity does not predict perfectly the precession in the orbit of Mercury – although it comes close. Thus, a TOE or a GUT (Grand Unification Theory) seems unlikely. There has been limited success in unification since only electro-magnetism and the weak force have supposedly been united. This desire for a Grand Unification Theory stems from the same psychology as Eastern philosophy's search for mystical oneness called "Brahmin" in Hinduism. *Star Wars* seems to borrow from both modern physics and Eastern mysticism in its famous line: "May the force be with you." And the force is said to connect all things in the universe.

So, to the social scientist, there is nothing strange or mystical about indeterminacy, uncertainty, randomness, or the observer effect. Particles are said to change their behavior when observed, and so do humans. Industrial psychologists have found that people are more productive on their jobs when they are conscious of being observed – no big surprise there.

Some social scientists believe in free will, and that would account for some unpredictability if free will is true. Other social scientists believe in total determination of behavior even to the point of believing that something determines the "will". There is some factor, they believe, that determines what we "will" to do. It is impossible to know which theory of human nature is correct because all the variables can never by specified, so there are hidden variables. The best we can say is that human nature lies somewhere between chaos and order. Certainly, we can predict a pattern of behavior for individuals and groups just as the quantum physicists can predict a pattern for an aggregate of particles. So human behavior is influenced but not totally determined by the factors we can know.

Yet, in modern physics, it is difficult to reconcile the idea of uncertainty, randomness and unpredictability with the most accurate clock known to man, namely the cesium clock which relies on quantum emissions. This quantum clock is said to deviate from perfect time by two nanoseconds per day - that's 2 billionths of a second per day. Then there are our electronic contraptions that we have become addicted to, from computers to smart phones. These devices have such a high reliability that the idea that they depend on some uncertainty principle with random fluctuations seems beyond credibility. And, if quantum phenomena are so unpredictable, how is it that chemical reactions involving electron sharing and exchange are so predictable? Perhaps a quantum physicist would say that randomness at the micro level averages out to predictability at the macro level. How such order could come out of chaos requires a mystical mindset.

CHAPTER 4: THE SCIENTIFIC METHOD, LANGUAGE AND THE PHILOSOPHY OF SCIENCE

Albert Einstein said: "Imagination is more important than knowledge. For knowledge is limited to all we now know and understand, while imagination embraces the entire world, and all there ever will be to know and understand." (Good Reads)

Feynman once said, 'Science is imagination in a straitjacket.' It is ironic that in the case of quantum mechanics, the people without the straitjackets are generally the nuts. (Brainy Quotes: Lawrence M. Krausss, theoretical physicist and cosmologist)

Classical, Relativity and Quantum Theory

Key Words: Critical Thinking, Epistemology and the Philosophy of Science

What is Science? Almost every science textbook, including social science texts, begins with the "Scientific Method" and attempts to define what science is and is not. The procedure described is monotonously the same regardless of field. The essence of the method can be summed up as rational-empiricism or logical-positivism where logic works hand in hand with empiricism, but empiricism is the final arbiter of the truth of a theory. Thus, science is different from pure math in which the proof of a theorem is based upon logic rather than measurement and observation. For example, the proof of a geometric theorem is based on logic because the inexactness of measurement makes it impossible to prove empirically.

Epistemology and Ways of Seeking Truth

This book is about the philosophy of science as it relates to modern physics. In this section, the book will deal with the history of the scientific revolution and the evolution of the scientific method. It will examine the scientific method and how it is different from other methods of pursuing truth. Let us begin with August Comte, a philosopher who is widely regarded as the father of scientific sociology (Martineau 1853). Now, to some, the idea that the study of people and society is a science is an oxymoron or self-contradiction. How can the study of people, as unpredictable as they are, be reduced to scientific laws? And, if one believes in free will, then human behavior is not determined and not subject to scientific laws. Obviously, Comte did not believe in free will and thought that human behavior is determined by laws in the same way that physical phenomena are determined by mathematical, physical laws. In arriving at the scientific revolution, Comte saw that society had evolved through three stages.

1) Theological mysticism – there is a transcendent truth that has to be revealed through spiritual means and is not comprehensible to the rational mind.
2) Metaphysics – truth is sought through pure logic and abstraction within the mind in the manner of Greek philosophers. For example, Geometry theorems have to be proven by logic, not observation.
3) Rational-Empiricism or Logical Positivism (science) – arriving at the truth through the scientific method of testing logically-derived, falsifiable hypotheses with empirical methods of rigorous observation and experiment.

While Comte saw society progressing in a linear fashion from mysticism to the scientific method modeling sociology after physics, he did not envision that some disciplines of science might come full circle back to mysticism and metaphysics. So, while he saw progress as a straight-line phenomenon, it appears that there may be a circular process. The adage "History repeats itself" may have some application here. Now, while physicists such as Capra, Bohm, Heisenberg and Bohr had the courage to embrace the fact that modern physics, in some ways, is more akin to Eastern mysticism than Western logic, many physicists vehemently deny this transition in thinking and are insulted at the suggestion that physics is anything but a hard science based on logic and rigorous experimentation and observation. In a small survey I performed in which I asked the question as to whether modern physics uses a different kind of logic than other sciences, all three respondents answered "no".

Yet the pioneers of quantum physics actually embraced Eastern mysticism as the only way to understand the counter-intuitive theories of quantum mechanics. Having taken Indian history in college, when I began reading quantum theory, I said, "This is very much like Hinduism and Buddhism." Then I found Fritjof Capra, and he confirmed everything I was thinking and more in his very engaging book, *The Tao of Physics*. I had been struck by the similarity between the Hindu saying: "I exist and don't exist," and the idea from quantum physics that a cat can be dead and alive at the same time if its life depends on a 50-50 quantum probability. The Hindu paradoxical saying also seemed to agree with the idea that a particle can be here and there at the same time. These are the kinds of concepts that Western philosophy cannot embrace because Western logic is dualistic, i.e., contradictions cannot coexist in the same theory. Western science is structured in such a way that hypotheses must be falsifiable, i.e., an hypothesis cannot be true and false at the same time; thus a cat cannot be dead and alive at the same time. Of course, allowing Eastern thinking into science can produce some conclusions that Western logic would never lead to. For example, in a superposition situation in quantum physics, the universe splits to actualize each possibility in the Many Worlds Theory. Surely, such as theory goes beyond the scientific method.

The Semantic Domains of Science: Physics and Metaphysics

It may seem pedestrian to re-examine the basic concepts of science, but I think the foundation of any structure should be inspected from time to time so it does not collapse of its own weight. There are five central concepts of science: matter, energy, space, time and consciousness. Let's view these on a continuum that progresses from the tangible and physical to the abstract and metaphysical.

As one progresses on this continuum from left to right, the concepts become less physical and more abstract and metaphysical. Here, I am using the term metaphysical to denote something abstract, inferred, or imagined rather than observed and empirical – in a word, Greek rationalism.

Now, in a sense all theories are metaphysical because they are composed of concepts, abstractions and representations of material reality. However, what makes a theory empirical is whether it is based upon observations in the physical world.

Time and space are relational concepts rather than direct observations that can be made with material objects. My primary concern here is with space and time, but I will write briefly about the nature of mass and energy. We have two ways of measuring the stuff we call matter: mass and weight. Mass, as measured by moles, amounts to the number of atoms made of protons, neutrons, electrons, and perhaps other subatomic particles that make up a particular piece of matter. To wit, 12 grams of carbon12 has $6.02214179 \times 10^{23}$ atoms in it. Weight, on the other hand, is the attempt to measure the amount of matter in a substance by how much gravity pulls it toward the center of a massive object such as the earth. Of course, the same mass will weigh more on the earth than on the moon and more on Jupiter than on the earth. Density is the amount of mass per unit of volume (space) that is measured. Again, Einstein confuses the language by saying that an object gains mass as it is accelerated and would obtain infinite mass if accelerated to the speed of light. The mass of an object (the number of protons, neutrons, electrons, etc.) does not increase with speed, but force (which acts like mass) does increase with speed. Newton was much clearer on this issue in saying that Force = Mass x Acceleration. Thus, we can say that while an object's mass does not change, its force in a particular direction can change with its acceleration, or we could say that the object's force changes with its momentum which is equal to its mass x velocity. Hence, contrary to Einstein, an object's mass has nothing to do with its speed or acceleration.

Now some physicists do not define mass as matter. An alternate definition is that mass is "resistance to acceleration". However, this definition of mass is the same as the definition of *inertia* which is also the resistance of a material object to acceleration whether the state of the object is inertial rest or inertial motion. As Newton said: "A body at rest or in motion continues at rest or in motion in a straight line unless acted upon by a force". This property of matter could be described as either *inertia* or *mass* in this alternate definition of mass. Nevertheless, the only thing known to resist acceleration is *matter*, so it could be said that mass defined in this way is a property of matter.

I believe that Einstein was right about the interchangeability of mass and energy – although some have argued that $E=MC^2$ can be derived without relativity by using classical concepts. I think of energy as high speed, fine particles of matter which carries various kinds of waves, and I think of matter as slower, condensed particles of energy which also has waves associated with it. Actually, energy which produces force is any matter in motion whether particulate or massive. Chemical energy produced by burning fossil fuel, creating heat and expanding gases, drives an automobile whose force is its mass times its acceleration. Thus, energy and force are manifested at the molecular level and the macro level.

Here again, we encounter contradictions and confusion in physics theory. **The gluons that supposedly hold quarks together in the nucleus of atoms are said to be massless.** To begin with, the idea of a massless particle is a contradiction in terms since any particle by definition has mass even if one is talking about a particle of energy. While the binding force of the nucleus is allegedly massless, the periodic table shows that the atomic mass of most elements is expressed, not as a whole number, but as decimal amount. For example, the atomic mass of Carbon is 12.011. Now there are at least two explanations as to why Carbon's atomic mass is not a whole number. One is that the atomic mass is the average of the atomic masses of all the isotopes of the element. The other is that the binding energy (gluons) also has a slight amount of mass and this is added to the protons and neutrons in the nucleus (University of Illinois Physics website 2007). If this second explanation is so, then the gluon cannot be massless – just as no particle of matter or energy can be massless. If it is massless, it doesn't exist. Only with Zen-type mysticism can something be massless. Perhaps one can say that vacuum or space is massless because these concepts represent the absence of matter and energy.

There are others in the physics community who argue that mass is not the amount of stuff you have, but is the resistance to acceleration (i.e., mass is inertia). Of course, what else is resistant to inertia besides mass and energy? But, they maintain that rest mass is different from mass in motion because mass in motion acquires additional mass. This assertion represents a confusion of mass, momentum and force. A piece of matter being driven to a high speed does not take on any additional protons, neutrons, electrons or binding energy. It may take on more force (Force = mass x acceleration) or more momentum (mass x velocity), but it does not add any new matter as it is being accelerated.

There are others in the physics community, however, who maintain, in the manner of Hinduism/Buddhism, that matter is somehow not real and that the material world is *maya* (illusion). Paul Davies combines his thoughts with Greene in commenting on the perhaps near-unreality of matter.

Apparently solid matter is revealed, on closer inspection, to be almost all empty space, and the particles of which matter is composed are themselves ghostly patterns of quantum energy, mere excitations of invisible quantum fields, or possibly vibrating loops of string living in a ten-dimensional space-time (Davies 2010: p. 65).

I suppose Davies is referring to Rutherford's experiments which involved shooting beta particles (electrons) through very thin gold foil (a few thousandths of a centimeter thick) to show that there is much empty space between the nuclei and electron shells of atoms. However, when Rutherford shot larger particles such as alpha particles (helium nuclei) at the foil, they tended to be absorbed at 100 times the rate of smaller beta particles. Now, what if Davies takes a piece of lead two inches thick and fires alpha and beta particles at it? Then we see that the atoms are packed together a little more tightly than in an ultra-thin piece of foil. And, what of the weight of the lead? Is that mostly empty space that is contributing to the mass of this heavy element? Further, the "ghostly patterns of quantum energy" are not "nothingness". After all, don't physicists agree that $E=MC^2$ and thus energy is equivalent to mass and contributes to the weight of lead and other elements. It really strains the intellect to believe that all this nothingness adds up to something as heavy as lead. An infinity of nothing will not add up to something. Furthermore, real science demands empirical proof, which means that to find support for a theory, you must produce **physical** proof that matches your hypothesis. If the physical world doesn't exist, then we might as well give up on the scientific method. How can physicists deny the very thing that would prove their theories?

Perhaps, rather than combining space and time into one entity (spacetime), we should combine matter and energy into "mattergy". It is interesting that these physicists, in the manner of Eastern mysticism, deny the existence of matter which is the stuff that physics was intended to study. There is no disembodied energy without matter. If I want to make an atomic bomb, I need some Uranium or other radioactive matter; if I want to generate electricity, I must have some magnetizable metals which I must move in relation to each other, and if I want to generate light, I must have something to carry electricity to a filament or gas that with cause it to release some photons. In summation, this book assumes that matter is real stuff and its behavior is the proper study of physics. Matter is the objective reality that provides empirical proof in the study of the physical world, and the study of the physical world is the business of physics.

Consciousness and Science Revisited

The ultimate purpose of science is to increase human knowledge of the universe and thus to expand our consciousness. In classical science, however, the scientist's consciousness does not directly affect what s/he is studying because material reality is seen as objective and has a separate existence apart from the human mind. It is assumed that the material world existed prior to humans arriving on the scene; therefore, the world does not depend on human perception. Not so in quantum physics where physicists believe that the observer has a direct effect on the outcome an experiment. The Copenhagen interpretation of quantum reality is that a particle's location and speed cannot be known simultaneously and that the particle's location can only be known when the experimenter makes an observation and collapses the wave. Not only did the experimenter not know where the particle was, but the particle did not have a definite place until the experimenter made it so with her consciousness. In a metaphorical sense, the particle didn't know where it was either. This sounds very much like New Age philosophers who contend that "you create your own reality." Those who follow the Copenhagen interpretation might argue that consciousness only affects reality at the quantum level and not on the macro level that New Agers are talking about. However, Erwin Schrödinger, who had classical leanings, imagined the famous cat box experiment where the cat's life depends on a 50-50 quantum probability (Gribbin 1984). Following the Copenhagen implications to their logical conclusion, Schrödinger says that the cat would be both dead and alive, or perhaps neither dead nor alive, until the experimenter makes an observation that collapses the wave.

While Schrodinger meant this thought experiment as a critique of quantum theory, his thought experiment is taken seriously by some quantum theorists who obviously believe that the consciousness of the observer actually affects material reality. In actuality, Schrödinger meant for his cat-box thought experiment to be a parody on the Copenhagen interpretation thus reducing it to absurdity. He was really saying that quantum theory has lapsed into mathematical mysticism in contending that the cat was both dead and alive until an observation was made. Although Schrodinger composed the famous equation which describes the quantum wave function, he later said about quantum physics: "I don't like it and I'm sorry I ever had anything to do with it" (Gribbin 1984: preface).

However, some physicists do not take this thought experiment as a parody but actually accept as reality that the cat is both dead and alive until the observer makes a measurement. Yakir Aharonov, who was a student of Bohm, contends that there are two probability waves, not one, that determine the hapless cat's fate. He says that there is one wave coming from the past and another wave that is coming backward from the future, and when the two meet in the present, they collapse to become material reality. Thus, the poor cat is caught in a cross-fire between future and past (Discovery Science Video 2002: *Uncertainty*).

Now let us examine in greater detail the three approaches to truth that August Comte identified: 1) mysticism, 2) metaphysics or rationalism, and 3) Empiricism or positivism. Let us determine which of these ways are within the scientific domain and which are not.

Mysticism: Mysticism, as I define it, is a type of knowing or believing that transcends human reason or understanding. As such, it is beyond language to define and must be "caught" by intuition rather than understood through logic. It is said not to be understandable by language because language breaks the unity of reality into categories or chunks to be communicated and thus fragments and distorts the mystical oneness of the cosmos. Apparent contradictions can coexist since they transcend the dualities of language categories. Since mysticism transcends language and logic, it is sometimes associated with supernatural forces or beings who have the ability to violate natural law as humans understand it. This kind of thinking was formerly not considered scientific because, if contradictions coexist in the same theory, then no hypothesis can be falsified. Since humans need mysticism, as Capra said, it seems to be an innate part of our psychology. There is a persistent belief among humans even in the scientific age, that there is a separate reality that transcends the physical reality brought to us by the senses. Some seek this alternative reality through the use of mind-altering substances that give us a different view of the world; others seek this other reality through meditation, religious ecstasy, and, no offense intended, some of my friends seek to know this other world through modern physics. There is no dearth of popular physics books that deal with the outer limits of the universe. Many of the readers are eager to know about a world where the laws of physics of the very small and the very large are so different that it "blow their minds." To some, the more outlandish and exotic the theory, the more interesting it is. Favorite topics are time travel, multiverses, quantum quandaries that defy logic, wormholes, black holes. I think Capra was right – humans need mysticism – even scientists who deny it. But has mysticism crept into science? I think so, and this book will offer evidence for this thesis.

Metaphysical rationalism: The prefix "meta" has several meanings including "beyond, after, or about" and often means "self-referential." Thus, meta-communication is communication about communication (for example, tone of voice or body language are implicit communication about explicit spoken communication). "Meta" can also mean beyond, so that metaphysics is often taken to mean something beyond the physical world suggesting something supernatural or spiritual. However, in this work, metaphysics will be taken to mean "about". So metaphysics is the mental representations we make *about* the physical world, rather than the physical world itself. This method of pursuing knowledge is through reason and logic. The ancient Greeks developed this approach to its zenith. Plato said that truth is in the mind discoverable by logic and that the exterior world of the senses is illusion and only a poor reflection of the perfect forms of the mind. Thus, a proof in geometry could not be determined by measurement or observation – that would only be an approximation. Geometric proofs must be discovered by logic and the perfect forms innate in the mind that must be awakened by reason. Hence metaphysics and rationalism are the basis of geometry and other forms of math. This method by itself is not science since it is based on the notion that the information gained from the senses is illusory. However, math and geometry along with qualitative language are useful tools in science in measuring and analyzing data gathered from the empirical realm. Nevertheless, math, geometry, and language are metaphysical entities that are superimposed upon the empirical world to aid in our understanding of it. They are not inherent in the physical world itself. They are in a word, "epiphenomena", i.e., superimposed on the natural world. Einstein's preference for metaphysical mathematics and thought experiments over the empirical is summed up in this statement:

The axiomatic basis of theoretical physics cannot be extracted from experience...Nature is the realization of the simplest conceivable mathematical ideas...we can discover by means of the purely mathematical construction of the concepts and laws connecting them to each other...Experience may suggest the appropriate mathematical concepts, but they most certainly cannot be deduced from them...But the creative principle resides in mathematics...I hold it true that pure thought can grasp reality, as the ancients (Greeks – my insertion) dreamed (Yaglom 1980: p. 187).

This kind of thinking is no doubt the basis for Einstein's many thought experiments or *gedanken,* which he apparently held as superior to empiricism through experimentation. In relativity, we see the beginning of the deviation from the empirical methods of Galileo and a move backward toward Greek metaphysical rationalism where truth is to be found purely in the mind. In quantum physics, we see this trend toward mathematical metaphysics taken to the limit.

Rational-Empiricism or logical positivism is based upon the idea that there is a real world out there that the senses inform us of. It makes the opposite assumption that rationalism makes in that it sees the empirical world as objectively real and that our senses do not, by and large, deceive us. When our senses deceive us, the logical mind connects the unseen dots, and when our logic does not match the empirical world (as with Aristotle and his 10 and 1-pound weights), our senses correct our reasoning. With our senses working with our minds, we make approximate representations of the world with the metaphysical tools of math and logic and our empirical observations. Hence, this approach combines the senses with the rationality of the mind to verify the truth. The rational mind generates hypothesis about the empirical world, but the senses (or instruments that extend our senses) test the accuracy of our hypotheses and are the final arbiter of what is true. This careful combination of mind and senses, then, is the stuff the scientific method is made of. One without the other is not science.

Zeno's paradox is the epitome of Greek rational idealism. Zeno was saying with his thought experiment that your senses deceive you, and that the mind with its logic can reveal the truth. Therefore, the observation that Achilles catches the tortoise is an illusion. The real truth is that he doesn't catch the turtle at all because logic says that every time Achilles catches up to where the turtle was, the turtle has moved on. Modern physics subscribes to the same kind of metaphysical rationalism in many areas. For example, we are told that our senses deceive us in creating the illusion that time flows one way. The rational truth, according to some physicists such as Brian Greene, is that time is a block where past, present and future are already laid down and therefore time does not flow at all. Physicists also tell us that when our kinesthetic senses make us feel "centrifugal force" when moving in a circle, that too is an illusion.

Galileo demonstrated the opposite. He showed that Aristotle's mind deceived him when his logic told him than an object weighing 10 times more than another object would fall 10 times faster. Galileo's approach to truth is said to be science and the beginnings of the scientific revolution.

Metaphysical vs. Mystical

In metaphysical rationalism, the senses deceive us.
In mysticism, the rational mind deceives us.

Mysticism is based on the idea that there is knowledge beyond the comprehension of the rational mind. That kind of knowledge is usually considered to be in the spiritual domain rather than the material realm. Mysticism is comprised of these elements: 1) Truth beyond the rational mind means contradictory things can both be true, e.g., Hindu proverb: I exist and don't exist and in quantum theory: a cat can be dead and alive at the same time, 2) Consciousness affects the physical world, e.g., the Copenhagen interpretation is that the observer's consciousness determines whether the cat is dead or alive and the participatory anthropic principle indicates that humans created the universe and its history with their minds, 3) Oneness or holism: everything is a part of a seamless whole and the notion of individuality or separateness is illusion and is the cause of conflict and pain, 4) The material world is maya or illusion.

As stated previously, there are many definitions of "metaphysical" and some of them include mysticism. For the purpose of this book, I will make a distinction between metaphysics and mysticism and will explain that difference here. By metaphysics, I mean pure logic and rationalism in the Greek sense; hence I call it *metaphysical rationalism* to distinguish it from other definitions of metaphysics. For example, one can imagine the expression of the square root of -1 in mathematics, but there is no counterpart in the physical world to this mathematical expression; therefore, it is purely metaphysical. I am thus using Immanuel Kant's definition of metaphysics:

"...*nothing more than the inventory of all that is given us by pure reason, systematically arranged* (Case 2013: Kindle Locations 49-50)."

Thus pure metaphysics alone is not physics – metaphysics is what we impose on the physical world with the mind or what we create with the mind that has no physical counterpart. Metaphysics (or logic) may generate hypotheses, but those hypotheses do not become physics until they are tested in the empirical world by experiment or careful observation. Mathematics, as well as qualitative language, are metaphysical in nature, but the belief that our senses (and mechanical extension of our senses) are a fairly accurate representation of physical reality is empiricism. Pure mathematics, not tested against empirical reality, is purely metaphysical and unscientific, but an hypothesis involving math and verbal description that can be tested against empiricism is within the scientific domain. For example, the idea of a multiverse (many universes) is a mathematical projection, but since other universes at this time cannot be observed, directly or indirectly, the multiverse hypothesis is not scientific and is purely metaphysical.

To distinguish metaphysics from mysticism and rational-empiricism (science), consider the following quote by Descartes and a proverb in Hinduism.

Mystical, non-rational thinking - Hindu proverb: "I exist and don't exist."
Metaphysical Rationalism – Descartes: "I think; therefore I exist."
Rational-Empiricism (science) – I am aware of myself, I see myself in a mirror, I hear myself speak, I feel my body; therefore I exist, and I believe my senses give me a fairly accurate representation of physical reality outside myself.

Descartes' statement is a metaphysical statement that is logical and rational. He sets up a contradiction and then shows that both statements cannot be true. If one is true, then the other must be false. Thus, it is a true-false, yes or no, test which leads to a *reductio ad absurdum* (a reduction to absurdity) of the false statement. Mysticism, on the other hand, sets up a contradiction and tries to break the initiate out of dualistic, rational thinking to transcend the contradiction and perceive oneness. The underlying belief is that separateness and duality are illusions and that the only reality is undifferentiated oneness that transcends plurality.

Descartes' syllogism goes like this: I don't exist. But in the act of doubting my existence, I am affirming myself as a conscious being capable of reflecting upon myself. Therefore, a conscious being cannot **not** exist. The contradiction is between non-existence and consciousness, and you can't be non-existent if you are conscious. Thus, the original statement "I don't exist" reduces to absurdity. You might doubt the existence of everything else, but you can't logically doubt your own existence - because you are a being that is conscious of doubting your existence – circular as that may sound.

The Hindu proverb is mystical because it defies Cartesian logic. It is non-logical because a contradiction is allowed to exist in the same statement – unlike Cartesian logic. Both are products of the mind, but one is rational and the other is non-rational and mystical. The point of this riddle in Hinduism is to force the practitioner of meditation to become conscious of the oneness that transcends dualities and the categories of language. Likewise, a Zen koan is a contradiction designed to confuse the rational mind and open up the intuitive mind to non-dualistic thought. Perhaps the best known koan given to an initiate is "What is the sound of one hand clapping?" So the contradiction that gives the mind pause is one hand vs. making a clapping sound. Of course, everyone knows it takes two hands to clap together and make a sound. Later we shall see the Fritjof Capra makes use of koans to explain concepts in quantum physics. Herein we see the divide between Western rationalism (largely championed by the Greeks) and Eastern mysticism. *In Western logic, contradictions cannot coexist; whereas in Eastern mysticism, they can coexist.*

Theravada Buddhism - the Ultimate Mysticism: 2 synthesized to 1 and 1 reduced to zero

Theravada Buddhism is perhaps the purest form of Buddhism in that it is thought to be the closest to what the Buddha taught originally. It involves several steps (jhanas) to achieve the ultimate nirvana. The purpose of meditation is to renounce attachments and to transcend the illusion of duality and separateness. Since language with its categories forces the mind into dualities, the aim is to synthesize dualities into a oneness (2 to 1, if you will) somewhat like Hegel's synthesis of thesis and anti-thesis into an integration. When oneness is attained, it is indivisible. In a mathematical sense, the number 1 cannot be divided and get 2 whole numbers. Hence, non-linguistic consciousness stops the flow of thought in which the meditator is aware of only consciousness, i.e., s/he is conscious of consciousness with no objects occupying consciousness since objects are separate, individual entities.

The final stage of renunciation of attachment comes when the meditator realizes that consciousness and the search for nirvana itself are attachments. Thus, consciousness creates another duality: desire for consciousness vs. desire for giving up all attachments, i.e., non-consciousness. Therefore, to create total renunciation of desire, the seeker must renounce consciousness entirely. In mathematical terms, this renunciation results in zero consciousness and 0 cannot be divided by 2 or any other number – zero is truly indivisible unlike 1 which can be divided into two halves.

The evidence that modern physicists believe that mysticism, rather than rational-empiricism, is the only way to comprehend the universe is summed up in these quotes:

Your theory is crazy, but it's not crazy enough to be true (Neils Bohr)
We must be clear that when it comes to atoms, language can be used only as in poetry. The poet, too, is not nearly so concerned with describing facts as with creating images and establishing mental connections (Neils Bohr). Bohr also told Einstein, when he objected to the idea of instant communication(faster-then-light) between entangled particles, "Mr. Einstein, you're being too logical."

The universe is not only queerer than you think, but queerer than you can think (J. B. S. Haldane).

By contrast, metaphysical concepts are born in pure logic. They are ideas or constructs that we impose on the world; thus, they are epiphenomena, rather than physical phenomena. They are representations we make of the world, not the world itself. The world is physical; the maps we make of it are metaphysical. Even though maps are made of materials, they had to begin in the mind and are therefore metaphysical in nature. Some metaphysical concepts may have a close approximation to reality, but others may be pure fabrications of the mind as in some geometric figure such as a dodecagon that you have probably never seen in nature. In the same way that early humans imposed a shape and design on rocks to form arrowheads and other tools, we impose theories on nature to try to understand it. As Richard Feynman once said: "Each piece, or part, of the whole of nature is always merely an approximation to the complete truth, or the complete truth so far as we know it. **In fact, everything we know is only some kind of approximation because we know that we do not know all the laws as yet**" (Brainy Quotes).

Metaphysical concepts are like the lines on the globe that represent the world. The earth is a real physical phenomenon (although some physicists would dispute even that), but the lines of longitude and latitude we draw on the globe are useful for navigation - still they are metaphysical properties we superimpose on the world. In other words, the sphere of the earth could have been described using 100 degrees or any other number base, but the founders of geometry arbitrarily came up with 360 degrees (founded on base 6 in mathematics). It works that the ancestors used 360, but it doesn't accord well with our decimal number system.

Geometry and other forms of math reside in the mind and have a life of their own outside the physical world. These metaphysical concepts can be used as tools to describe the physical world, but they are not the physical world themselves. To fail to distinguish this is to invite error. The point is, in forming a theory, if you don't recognize an objective, physical world that is separate and independent of our consciousness and our perception of the world, you get some really absurd theories such as the ones I have described. In other words, almost anything goes. Paul Davies indicates that physics is turning more and more away from empiricism to metaphysics. He says that "The history of physics is one of successive abstractions from daily experience and common sense, into a counterintuitive realm of mathematical forms and relationships" (2010:p. 65).

A good example of what Davies is referring to is something mentioned earlier, namely, String Theory which posits that there are 11 dimensions (not just the three we operate in daily), and that elemental matter is made up of tiny 1-D strings that vibrate in 11 dimensions. The various frequencies of vibration create the elementary particles (quarks, protons, neutrons, electrons, etc.).

Of course, the idea of 11 dimensions and matter existing in one dimension is purely metaphysical (a mathematical construct), because no one has ever observed or measured any dimensions other than the 3 we experience every day. And, there is no evidence that matter, no matter how small, can exist in one or two dimensions. Matter is 3D, and no matter how fine you cut it, it will always be 3D. It's like the riddle of the frog progressively jumping half way to the wall with each jump. The frog never gets there. So it is with matter, you can half it as many times as you like, but you never get to a 1 dimensional piece of it. Protons, electrons, neutrons and quarks are three dimensional. How could one or two-dimensional pieces of matter ever add up 3-D matter? So String Theory does not distinguish between physical phenomena based on observation and measurement - and metaphysical phenomena hatched in the brain. String theory may produce some pure mathematics (metaphysics), but physicists are finding that it does not describe any physical system, i.e., matter and energy. It seems that if string theorists had been thinking scientifically, rather than mathematically and metaphysically, to start with, they could have saved themselves a lot of wasted effort. Herein, we see the trend in modern physics toward metaphysics and later we will see a trend toward mysticism. It is interesting that **physics** would come to the point of doubting the existence of the **physical** world which is the proper subject of its study.

One of the distinctions between metaphysics and mysticism is that in metaphysical logic, contradictions cannot coexist, but in mysticism contradictions can coexist. Therefore, we find that relativity is based more in metaphysics, and quantum theory is based more in mysticism. In relativity, Einstein engages in rather pure thought experiments which are metaphysical. However, in the thought experiments, contradictions cannot coexist. Therefore, in special relativity, he deduces that the speed of light cannot be both relative and constant and since he assumed it is constant in his original proposition, the speed of light is therefore constant and cannot be relative. This is proof by contradiction or *reductio ad absurdum* of the competing argument. However, in quantum mechanics, a superposition of probability means that a particle can exist in two or more contradictory states at the same time. Thus, a particle can be here and there at the same time or a cat can be dead and alive at the same time until an observation is made. Thus, contradictions can coexist and therefore the theory is based in mysticism by any reasonable definition of mysticism.

Furthermore, quantum theory claims that physical reality is particulate rather than continuous and whole. For example, energy is not continuous, but comes in small, discrete packets called quanta. And, matter is not continuous but made of atoms and subatomic particle that are discrete and detachable from the atom. On the other hand, some quantum theorists argue that reality is an undifferentiated whole that cannot be broken into separate, individual parts. Again, the mystic can have things both ways since the mystic is not bothered by contradictions.

> Nutshell: In one sense, all human knowledge is metaphysical because the sensory images our brain makes of the physical world are not the physical world itself, AND, the way we describe the world with language and mathematics is a representation – the map is not the territory. However, there are *pure metaphysics* and *metaphysical representations* made by observation which are accurate enough to enable our survival. An example of pure metaphysics is mathematical representations of extra dimensions which are unobservable or without indirect evidence of their existence. Such extra dimensions are pure products of the mind even as the square root of -1 is a pure mental construct. On the other hand, the calculation of the period of a pendulum is metaphysical in that it is processed in the mind, but the observation of a pendulum's swing is a representation of a physical reality if we believe our senses give us a reasonable representation of the physical world outside ourselves. If our senses do not give us a reasonable facsimile of the world, then we wouldn't have survived as a species.

Following up on the statement that all knowledge (even so-called empirical knowledge) is ultimately metaphysical in that it must be input and processed in the brain, let us look at the process of taking information into the mind and organizing it into theories. The process can be seen as occurring in three stages.

Sensation → Perception → Cognition

Sensation is the raw data that we bring into the brain with our senses such as sight and hearing, and perhaps the information that we bring into the brain by instruments that extend our senses such as telescopes, microscopes, Geiger counters, particle colliders, etc. Even with sensation, there is some summarization and filtering of the data that is allowed into the brain, but this is as close to pure empiricism as we can get.

Perception is the beginning of the coding, categorizing and interpretation of information that is brought into the senses. The way we perceive color is largely determined by our culture. Navaho Indians combine blue and green into one category, but English speakers separate these two colors into different categories. Hence perceptions are one step removed from empiricism and are more metaphysical in nature.

Cognition is the higher order, cortical thinking that is demanded in science and is two steps removed from sensation. It involves finding the linkages among categories – for example the link between falling objects and gravity. Even though some philosophers don't like to speak of causation, the ultimate goal of theories based on higher cognition is to find, not only linkages, but the causes of observed effects or a chain of causation that leads from one effect to another. Although cognition is two steps removed from the raw data of empiricism, the theories that emerge from it must match the sensations garnered from the physical world. However, competing theorists often claim that their theories match the empirical world better than the competition. Therefore, additional tests and more careful observations may have to be made to determine which theory is closer to the empirical world.

Steps in the Deductive Method of Science

The inductive method proceeds from specific to general, that is, it begins with specific facts derived from observation and attempts to generate general theories from those facts. The deductive method, on the other hand, proceeds from general to specific, that is, it starts with a general hypothesis that explains a phenomenon, and then attempts to find specific facts that will falsify the hypothesis or support it. Thus, the inductive method starts with empiricism, and the deductive method begins with rationalism. As can be imagined, most scientific research is deductive because inductive research is time consuming and expensive. The Achilles heel of deductive research is bias. It is understandable that a scientist has an emotional attachment to his/her theory and is intent on proving the hypotheses derived from it. We shall see several examples of "deductive bias" in reviewing scientific research in this book.

Cox and Forshaw (2010) in their book, *Why Does $E=MC^2$* outlines Richard Feynman's version of the scientific process.

First we guess at it (a new law in science). Then we compute the consequences of the guess to see what would be implied if this law that we guessed is right. Then we compare the result of the computation to nature, with experiment or experience, compare it directly with observation, to see if it works. If it disagrees with experiment, it is wrong…It does not make any difference how beautiful your guess is. It does not make any difference how smart you are, who made the guess, or what his name is – if it disagrees with experiment, it is wrong. That's all there is to it (Feynman Lecture 1964).

What Feynman leaves out is that the data obtained from an experiment may be interpreted in several different ways depending on one's theoretical or philosophical orientation. For example, the Michelson-Gale experiment that we shall discuss later was interpreted as support for the "ether theory" and the "theory of General Relativity" which contradict each other. Be that as it may, do physicists follow this scientific method to the empirical stage as laid out by Feynman? Feynman is quoted as saying that "science is imagination in a strait jacket. Lawrence Krauss added to that: "It is ironic that in the case of quantum mechanics, the people without the straitjackets are generally the nuts" (Brainy Quotes). While I wouldn't use the term "nuts" to describe theorists who shed the straightjacket or discipline of science, it is my perception that in some theories, such as string theory, mathematical proof substitutes for empirical proof and many hypotheses cannot be tested and thus go beyond the current domain of science.

In his lecture on the scientific method, Feynman is talking about the deductive approach to science. The following are the specific steps used in the deductive method in which the process begins with an idea as to what the answer is before putting it to the test. In the inductive method, the researcher ideally approaches the subject without theoretical preconception and simply investigates the phenomena.

Theory: In the deductive method, there may be an existing theory that attempts to explain the phenomenon in question. Although some physicists of mystical bent reject the idea of causation, the ultimate purpose of a theory is to find the force that causes a certain phenomenon. There are four major forces or causes that physicists have identified in nature, and the dream of physics is to find the prime mover of all forces by unifying all these forces and finding the root cause of everything. When a theory is supported by evidence beyond reasonable doubt, and no competing theory can explain the phenomenon in question, then the theory has become a law. Since there are very few laws, we are stuck with incomplete theories as Feynman has stated. Thus, a theory is like a blinder on a horse that prevents him from seeing things to his right or left and getting spooked. Theory thus directs our attention to specific variables that the theorist thinks are significant causes. A theory can also be thought of as a filter which allows certain frequencies of light in and blocks out others. Thus, a theory enables us to see some aspects of reality and blinds us to others because, again, theories are approximations – they are not laws. Many times, something is stated as a law of nature, when it should be called a theory of physics – big difference! The following are the steps of the deductive method.

1) Question: The method begins with a question about some phenomena. Ultimately, the scientist wants to know the cause or causes that produce a certain effect – although some philosophers of science reject the idea of causation and prefer to use the terms "independent variable" and "dependent variables" which amount to about same thing as cause and effect.

2) Hypothesis: In the deductive method, the scientist forms an hypothesis or "educated guess" as to the cause or the independent variable related to the dependent variable. At this point, logic is used to determine the most probable cause and to formulate an hypothesis that can be **falsified** as Karl Popper has said. Sometimes, a null hypothesis is formulated stating that there is no relationship between the suspected independent variable(s) and the dependent variables. For example, a null hypothesis might read that there is no relationship between taking vitamin C and the frequency of contracting colds or flu. This step is based largely on rationalism, i.e., what seems logical and reasonable. If evidence is found supporting this hypothesis, then it gives support to the larger theory of which it is a part.

An example of a non-falsifiable hypothesis is the prediction, made facetiously by Schrodinger in his famous cat-box thought experiment. Mocking the Copenhagen group, he predicts, based on the eponymous Copenhagen interpretation, that the cat is both dead alive when its fate is dependent on quantum probability. Although Schrodinger meant this thought experiment as a parody and *reductio ad absurdum*, some physicists still take it literally. The prediction is unfalsifiable because a dead-and-alive cat cannot be observed because no one would know what such an animal looks like. If the animal is dead, then it is not alive and if it is alive, it is not dead, so no observation can show that it is dead and alive simultaneously. When a theory is taken to its logical conclusion and makes a prediction that cannot be falsified, then that theory is unscientific.

Forming an hypothesis is where mysticism separates from science. In mysticism, contradictions can coexist and in science contradictions cannot coexist. If contradictions can coexist, then there is no falsifiability of the hypothesis since contradictory things can both be true in mysticism. For example, the idea that you can exist and not exist at the same time is mysticism, not science. Physicists tell us that the contradictions of quantum mysticism transcend rational understanding. If so, these contradictions transcend science. However, physicists use what is called "proof by contradiction" (where if one thing is true, the other cannot be true) when it is convenient and resort to mysticism (opposite contradictory things can be true) when convenient. For example, Einstein uses "proof by contradiction" in his thought experiment on special relativity. He shows if the speed of light is constant, then it cannot also be relative, and since the speed of light is constant, a space ship traveling at a high rate of speed with a headlight cannot add to the speed of light. Furthermore, Bell, in showing that there is action at a distance between entangled particles, argues that if such particles can communicate instantaneously at a distance, then this effect cannot also be locally caused. If instantaneous action at a distance is true, then the idea of local causation is reduced to absurdity. Yet, we are told not to use logic when accepting the idea that a cat can be dead and alive simultaneously and that a particle can be here and there at the same time.

3) Data Collection by observation or experiment: The scientist gathers data that is pertinent to the hypotheses and can either support or negate the hypothesis. This is where the **empirical** aspect of the method enters the procedure and it could involve structured careful observation, an experiment or a use of some instrument to extend the senses. Of course, if one doesn't believe that the material world is real (as some quantum theorists have said), then all the empirical evidence that one gathers is mere illusion. Science, however, is founded upon the belief that the physical world is real. Those who say it is not real are engaging in metaphysics or mysticism, not physics.

4) Analysis: The scientist analyzes the data to determine if the hypothesis is supported or not. This is where logic kicks in again in linking the data to the hypothesis. Mathematics are generally used as a tool to analyze the data and show trends or regularities through the use of statistics or calculus. With mathematics and qualitative language, inferences are made as to how well the hypothesis fits the data gathered and contributes toward a more general theory which might involve several related hypotheses. Although logic is used, in proposing hypotheses and analyzing the data, the ultimate arbiter of whether the hypothesis is true or false is, as Feynman says, the **empirical data** derived from experiment or careful observation. Hence in the philosophy of science, the scientific method is based upon rational-empiricism or logical positivism. What has been described here is the **classical approach** to physics which is based on materialism and empiricism. In **modern physics**, however, some theories indicate that the empirical, material world is an illusion and cannot be trusted - our senses deceive us, and truth must be found in mathematics and pure reason. So, in some way, philosophy has come full circle from Greek rationalism where truth is to be found in the mind, not the senses which are illusory, to the rational-empiricism of classical science and back to the Greek ideal of pure reasoning through mathematics which has become more the *modus operandi* of modern physics.

To illustrate this cycle, let's take the familiar Aristotelian vs. Galilean approaches to describe falling objects. Almost everyone knows that Aristotle, using the methods of Greek rationalism, said that an object weighing ten times more than another object will fall to the ground ten times faster. This logic seems plausible and he used mathematics to help prove his point, i.e., the parallel of 10 times the weight will produce 10 times the speed.

Now enter Galileo a few centuries later who applied rationality and **empiricism** to this question. He raised the question about what happens if one ties a ten-pound weight to a one-pound weight. Will they fall at the same rate as the one-pound weight or the ten-pound weight; will they fall at the average speed of the two; will the one pound weight retard the speed of the 10 pound one; or will the 10 pounder increase the speed of the one pounder? However, Galileo, unlike Aristotle, did not stop with reasonable speculation. He is said to have dropped a one-pound weight and a ten-pound weight from the leaning tower of Pisa and found that they fell at the same rate. Whether this is an apocryphal story or true is not certain, but what we do know is that he rolled balls of different weights down inclined planes and measured the time it took them to reach the bottom. He thus proved the principle of equivalence not by logic and reason alone, but by empiricism and experiment. Galileo apparently believed that this physical world was real and not a holographic projection – and that the empirical world brought to us by the senses was the final judge of the truth of an idea.

The following chart shows the various levels of the deductive method in formulating theory. If there are glitches in a theory, one should return to the base and evaluate assumptions and concepts.

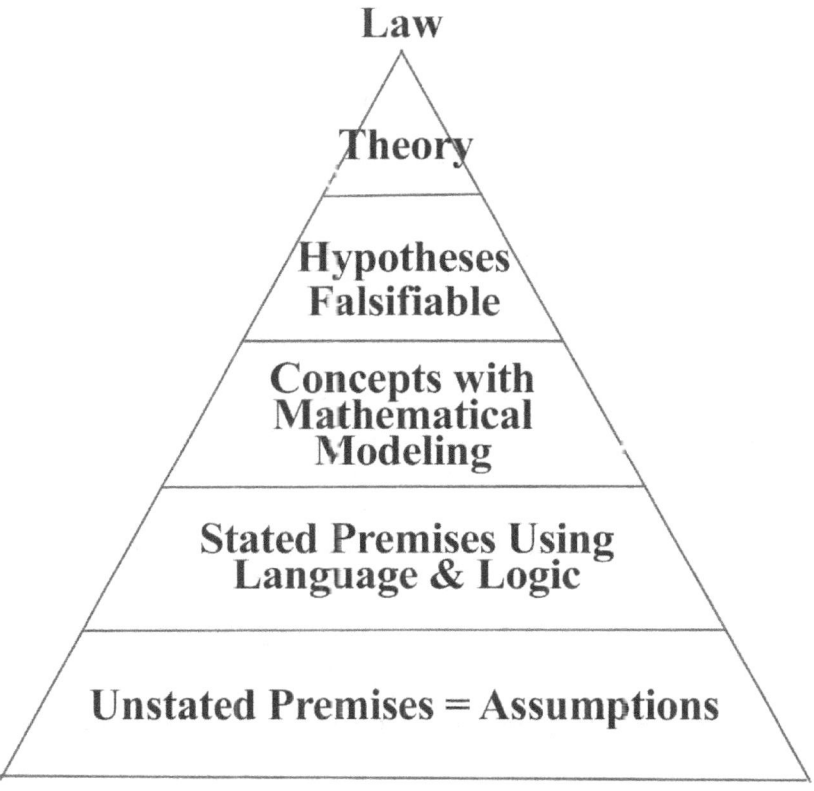

Karl Popper and Falsification
Karl Popper, famous philosopher of science, indicated that for a theory to be scientific, it must be falsifiable. That is, the theory must be structured in such a way that empirical evidence can contradict it. If the theory will accommodate contradictory evidence, then it is not scientific by Popper's definition. Popper went on to say that it may not be possible to prove that a scientific theory is true, but it is possible to prove that it is false. In other words, doing science is a process of elimination. If the scientist can eliminate a number of theories by proving them false, then that gives credibility to a theory which has not been falsified even though it is structured in such a way that it could be falsified. If Popper's thesis is applied to the Many Words Theory, it would be found to be unscientific because it cannot be falsified since it indicates that "decoherence" makes it impossible to see the new universes being created. String Theory would also be unscientific unless extra dimensions beyond the three we experience every day can be empirically demonstrated. String theorists say that unless they can find supersymmetric particles, the theory cannot be supported. However, even more basic than that, the theory cannot be supported unless

extra dimensions can be validated as a physical reality and not a mere mathematical projection. Another excellent example of a Ptolemaic type hedge to insulate a theory from falsification comes from the inflation or expansion theory of cosmology. It is said that the galactic clusters are moving away from each other faster than light. When the objection is raised that nothing can go faster than light, a hedge against falsification is erected. The hedge is that it is space that is expanding faster than light and that the galaxies are stationary relative to space. Since they are just along for the ride, their being carried by space doesn't count as motion. Kaku puts the cap on it: "Space is nothing and nothing can go faster than light". In other contexts, however, physicists say that space is actually something, not nothing. Is this a hedge or rational theory? I'll let you, the reader, decide.

> *The key to developing a good theory is that the theory must be modified to fit the facts rather than modifying the facts to fit the theory. Even when modifying a theory, one must be careful not to create epicycle-type factors to prop up a theory that is spurious.*

Metaphysical Rationalism in Modern Physics

Now in modern physics there are some who apparently revert to the Greek rationalist way of thinking that pure logic, not observation, is the path to truth. Einstein himself said that imagination is more important than knowledge, and most of his work involved thought experiments which are exercises in logic rather than observation. Let's take the notion of **block time** as opposed to **one-way flow of time** as an example. Brian Greene (1999) admits that all of human experience tells us that time flows one way from past to present to future, but, he goes on to say that our senses and perceptions deceive us and that all time (past, present, future) exists together. He clings to the point despite the fact that all physics experiments so far as I know are conducted with the sequence of time flowing in one direction, and there is no empirical evidence of time flowing in any other direction than from past to present to future. But what good is empirical evidence if our senses deceive us and the empirical world is not real – just shades and shadows of the perfect reality in the mind.

The example referred to earlier is **string theory** that is based largely on pure mathematical reasoning which requires eleven dimensions in order for the math to work – despite the fact that there is no empirical evidence for any dimensions other than the three we live in every day. Again, what good is empirical evidence when mathematical reasoning yields Platonically beautiful and elegant theories that messy empirical reality can never match? However, in describing the physical world, the math can be no better than the premises (assumptions) and inferences that it is based upon. Math, as a way of knowing the world, is imperfect just as empiricism is imperfect. Let's take the game of chess mentioned earlier. Elegant mathematics can be used to describe chess moves, and a computer program has been designed that beat the chess champion of the world, Gary Casparov. But, does the mathematics of chess describe any system in the physical world? I don't think so. The mathematics of chess is internally consistent because of the rules created in the mind. Chess math is self-referential and thus describes chess itself and nothing else. Does string theory math describe any physical system, or does it merely describe itself?

It appears that some physicists have come to regard math with Pythagorean reverence and a mathematical proof is regarded as a physical proof. We know that math, like qualitative language, is imperfect in describing the world. Take the circle, one of the most basic figures in all of geometry. The relationship between the diameter and the circumference (symbolized by the Greek letter pi) has been proven mathematically to be transcendental and irrational. Thus, the math says the circle cannot be squared. However, we know that a compass, which forms a radius, can be rotated 360-degrees to form a circle, and although we cannot draw a perfect circle, we know logically that "rotating the radius forms a circle" is true in the ideal, Platonic sense. Likewise, the circumference cannot be measured exactly, but using Greek geometry, the circumference can be cut in half (and any power of two), thirds, fifths and sixths. These operations would not be possible if the circle were truly transcendental in the geometric sense as the math says it is. Yet, how many physics formulas contain the illusive pi. When physicists speak of the mathematical laws of nature and use pi to formulate those laws, we know that they are dealing with an approximation rather than an exact, definable number.

Michio Kaku (2013) provides us with a perfect example of how modern physics has become almost pure mathematics, or in a word, metaphysics. He says that around 1905 math and physics parted company with physics moving in the direction of relativity and math moving into the domain of differential topology and hyper dimensions. He goes on to state that mathematicians believed that they had found an area of mathematics that had no counterpart in physical reality, and that mathematicians have a strong affinity for this pure, rarified math. In other words, the field had become pure mathematical metaphysics. However, Kaku continues that physicist then discovered string theory which involves the vibration of strings in 11 or 12 dimensions of hyperspace with supersymmetries. Thus, physics has created a new kind of math that supposedly describes physical reality and again unites math with physics. Kaku's final conclusion is that God should be conceived as a mathematician whose vibrating strings produce the music of the universe. What this trend really shows is that math and physics have not merged, but that math has taken over physics and made it into nearly pure metaphysics. Mathematical proofs have largely replaced physical proofs; therefore, modern physics is describing the workings of the mind, not the workings of the universe.

> Controlled observation is the foundation of the scientific method. However, there is **active and passive** observation. The Copenhagen interpretation of quantum physics asserts that the observer cannot be separated from the observed and thus consciousness emerges as a determinant of the outcome of an experiment. However, in order to determine the location of a particle, the experimenter must affect the particle in some way (such as hitting it with light) since it cannot be passively observed. Thus, the natural behavior of the particle has been changed by this active observation. On the other hand, much macro phenomena can be observed passively, i.e., without interfering with the object observed. In social science active observation would be watching a group of people who are conscious of your presence; whereas a hidden camera would represent passive observation. This important distinction is not made in quantum physics.

Mathematical Proofs vs. Scientific Proofs

"All structures that exist mathematically exist also physically." Max Tegmark

A mathematical proof is based purely on logic and rationalism; while a scientific proof is based on rational-empiricism. For example, the Pythagorean Theorem cannot be proven by measurement because any measurement will be inexact. It stands on the logic just as the bisection of an angle stands upon logic rather than measurement. Measurement, based on empirical observation, can approximate Platonic ideal forms such as a perfect triangle or a perfect circle, but it can never achieve perfection. Thus, mathematics is **metaphysics**, i.e., it is "meta" or "beyond" physics. Math is the epiphenomena that we superimpose on physical reality - it is like the lines of latitude and longitude drawn on the globe. Although these lines are not found on the earth, they help us to understand the shape of the earth and enable us to navigate it effectively.

Many concepts that define modern physics are derived from mathematical metaphysics rather than observation. For example, **extra dimensions** (besides the three we live in) come out of metaphysical assumptions, and there is the assumption that matter can exist in one-dimensional or two-dimensional space. I have yet to find a physicist or mathematician that can offer any evidence of extra dimensions or evidence for two-dimensional dogs; yet, as we have seen, string theory rests on the assumption of 11 such unseen dimensions and that strings of matter can exist in one dimensional space.

The use of zero in modern physics is a perfect example of how math is misused in theory and leads to false conclusions. As we shall see later in the Cosmology section, Krauss and Kaku argue that since most energies are dualistic and equal they cancel out to zero. Therefore, they argue, that the negative charges in the universe are equal to the positive charges of the universe, the net charge is zero. From the balancing of charges to zero, they argue that the universe could have come from nothing (i.e. zero). This non-sequiter will be addressed in a later work.

Likewise, **block time** is based on metaphysical assumptions and denies the empirical reality that time flows in one direction. There is not one empirical test that I am aware of that supports the notion of block time. Furthermore, the wedding of space and time, i.e., spacetime, which underpins the theory of relativity is an assumption that cannot be demonstrated empirically – it is an assumption that makes a theory work. Even Einstein himself initially rejected the notion of spacetime, posited by his professor, Hermann Minkowski. Einstein initially described "spacetime" as just so much sophistry. Later, when he found that it would make his theory of relativity work, he embraced the concept with great enthusiasm. As we shall see in later sections of the book, space and time are fundamentally different. An object can move in any of three dimensions of space, but, despite metaphysical arguments to the contrary, objects move in only one direction in time. Thus to critique any scientific theory, the first question to ask is "What are the assumptions that it is predicated upon."

Can Contradictions Coexist in Modern Physics as in Eastern Thought?

The idea that a particle can be both here and there at the same time is very similar to the Hindu concept that "I exist and do not exist" at the same time. Hindu thought is largely based on transcending dualities which are mutually exclusive. Hindus and Buddhists also see physical phenomena as "maya" or illusion - much like the Platonic concept that reality is in the mind, and the world of the senses is imperfect and therefore unreal. The concept of a **"massless particle"** such as photon would seem to be a contradiction in terms to the logical mind. How can something be a particle of matter and not have mass? Even if you say that a particle is a packet of energy rather than matter, energy is said to be equivalent to mass in Einstein's $E=MC^2$ and by the first law of thermodynamics that says that matter and energy are interchangeable. When asked the question of whether photons have mass, one of the physics professors who responded to my survey answered that photons do have mass because they have energy. The idea of a massless particle is also like Schrodinger's thought experiment where the cat is both dead and again resonates with the Hindu saying: "I exist and don't exist."

The concept of "**negative energy**" is another paradox, if not an outright contradiction in terms. Either there is energy or there is no energy. Negative numbers are imaginary and correspond to a social convention. For example, if one's liabilities exceed one's assets, then s/he is in debt and can be said to have a negative net worth; however, it does not seem reasonable that nature has debts. For instance, a thermometer measures temperature, and using the Fahrenheit and Celsius scales the temperature can go below zero into negative numbers; however, the negative numbers do not indicate total lack of heat. Only when the measurement of temperature reaches absolute zero on the Kelvin scale can there be no temperature or molecular motion. At that point, there is no below-zero into negative numbers. You either have heat or you don't.

"**Borrowing energy from the vacuum**" is another seeming contradiction. A vacuum is defined as the absence of matter and energy. Thus, borrowing energy from nothingness sounds like trying to borrow money from a bank that is broke and has no money. Quantum physicists say that particles are constantly popping up in space and disappearing. I would question how it could be possibly known that the particles that appear to be coming into existence did not have a previous existence. After all QM physicists tell us that waves are collapsed into particles by observation and the universe is saturated with light (photons), cosmic microwave background radiation, and particles from super novae.

These topics will be explored in greater detail later in this book, but for the purpose of this discussion of the scientific method and its relation to philosophy, the conclusion I draw is that classical science is rooted in the rational-empiricist tradition where metaphysics is combined with physics but where empirical reality is the final arbiter. Modern physics, I and others believe, has drifted toward almost pure metaphysics and has become decidedly anti-materialistic and anti-empirical. Thus modern physics is based more in Eastern mystical philosophy and classical physics is rooted in Western logic.

Consider these general analogies that distinguish between metaphysics and physics:

Metaphysics is to Physics as
Logic is to Observation
Language is to the Senses
Mind is to Matter
Subjective is to Objective
Eastern logic is to Western logic
Modern Physics is to Classical Physics

In the view of this author much of modern physics is anti-scientific because some theories deny the reality of the physical world, and science demands physical evidence to support theory.

Furthermore many theories, particularly in quantum mechanics cannot be falsified; therefore they are unscientific. These theories cannot be falsified because it is not possible to produce physical evidence relating to the theory. The Many-Worlds theory and String Theory fall into this category of non-falsifiability.

The following is a chart that shows the cyclical history of philosophy as it relates to science. Philosophy has made the full circle from mysticism to metaphysics to empiricism and back to metaphysics/mysticism.

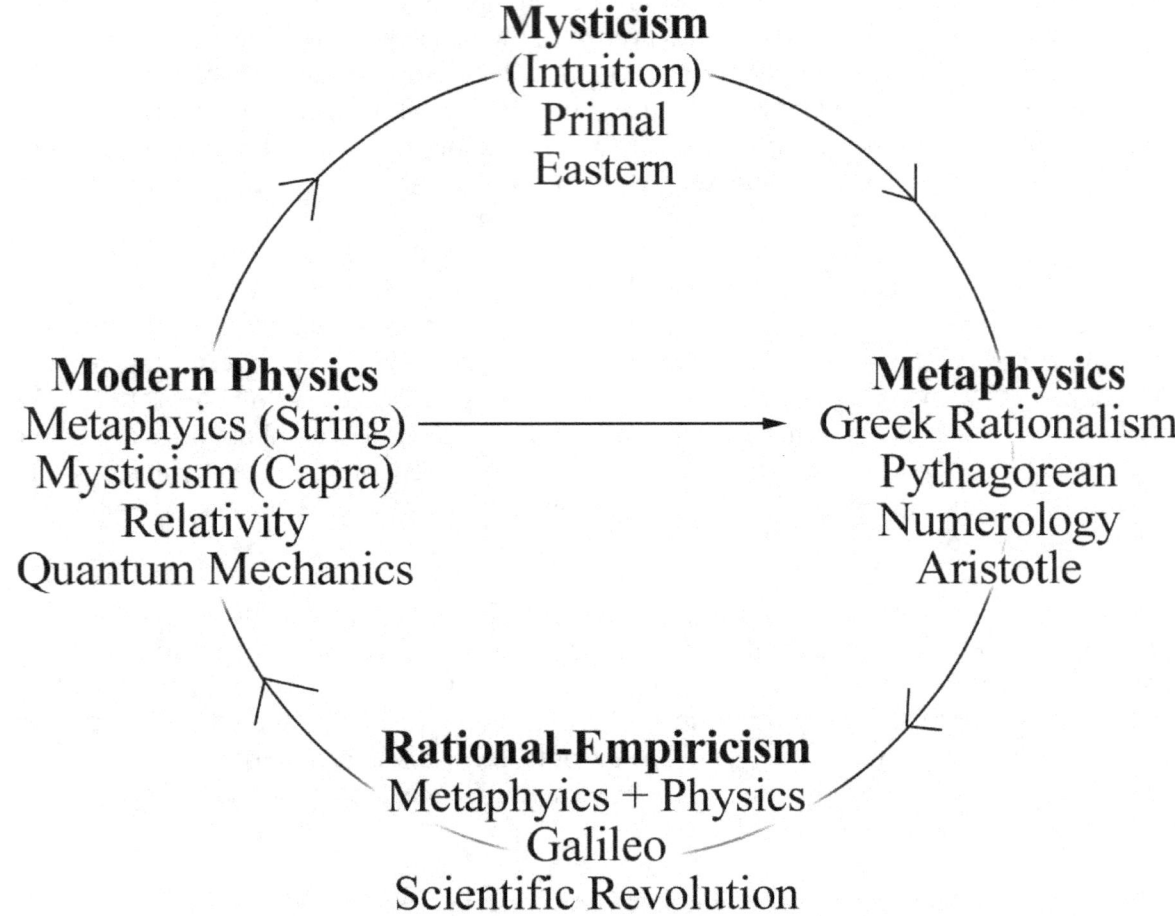

Nutshell: Metaphysical abstractions, such as spacetime, are not able to act on or be acted upon by physical forces. Only physical objects are subject to physical forces, to wit, the lines on a globe do not cause the earth to spin. Metaphysics describes and organizes our thoughts on physical reality – metaphysical representations are not physical reality in themselves.

Is Modern Physics Really Science in the Classical Sense?
1>Classical science is based upon the assumption of mind-independent, objective reality. For example, Galileo's experiments demonstrated by empirical observation that Aristotle's thought experiment about falling bodies was wrong. Galileo's observation did not cause all bodies

> regardless of weight to fall at the same speed. They fall at the same speed whether Galileo is watching or not.
> 2>On the other hand, in the Copenhagen interpretation of quantum mechanics, all reality is mind-dependent, i.e., the observation of the experimenter determines the outcome of an experiment. For example, the cat is both dead and alive until an observation is made.
> 3>Special Relativity leads to the twins paradox which means that the twin at rest and the twin in inertial motion cannot tell who is moving and who is not. Since each twin appears to the other as being in motion, each twin is seen as being younger than his counterpart because time supposedly slows down for objects in motion. In the scientific method, falsification is based on the idea that one hypothesis or the other must be true – both cannot be true. As a matter of fact, Einstein used this method in his thought experiments to show that if one proposition is true, then its opposite cannot also be true – in other words, proof by contradiction. For example, it cannot be true that the speed of light is constant in all frames *and* that a moving body emitting light will add to light's speed. Since both cannot be true, then the only truth is that the speed of light is constant. Likewise, if each twin can be younger than the other, then there is no falsifiability, and reality is not only relative but subjective and *appearance* is reality. The fallacy comes when the perceptions of the twins take precedence over the objective reality that Einstein states in the beginning, namely, that one twin is at rest and the other is in a constant rate of motion.
> * In neither of these cases does Modern Physics meet the standard of the Scientific Method.

The Quantum Physics, Mysticism, Spiritualism and Religion Connection

It is ironic that some physicists admit to embracing mysticism but claim to be atheists. Bohm and Capra, while embracing Eastern mysticism, also embrace atheism. I suppose that their model of mysticism is original or pure Buddhism (Theravada) which is said to be atheistic since it does not assert a belief in a creator god, and karma is seen as a natural, impersonal force like gravity. The fact that one has to suffer because of his bad behavior is not due to punishment by a personal god – the suffering is caused by an impersonal law of nature. Of course, a later form of popular Buddhism (Mahayana Buddhism) does indeed include religious elements such as beliefs in demi-gods and all manner of spiritual beings. However, I would disagree that original Theravada Buddhism is not religious at all. Even Theravada Buddhists believe in a soul or spirit that is not dependent upon a physical body for existence, and this soul can be reincarnated in many forms. This certainly is a belief in spiritual existence which is the essential ingredient in spiritualism and religion.

I have observed that atheist physicists who embrace Theravada, also embrace Darwinian evolution which explains the origin and development of life as based on pure chance acted upon by an environment that selects for adaptive traits and rejects non-adaptive ones. Some of these same physicists also believe that the origin of the universe, namely the Big Bang, came about by chance. Other physicists who have an atheist orientation believe in evolution but that it is highly improbable that it could have come about by chance interaction of particles. These physicists see that the creative order is innate in nature and does not depend on some superordinate being who is separate from the creation process. Still others embrace a kind of intelligent design in which the idea of a separate, transcendent being is implied or stated. Bob Berman, science writer, says it well (2017: p. 10):

A third choice (other than intelligent design and random causation), embraced in the East, is that nature itself is smart and exhibits its own form of intelligent design...an awesome underlying intelligence in nature seems obvious...my hope is that the subject of abiogenesis on Earth or on exo-planets leads us to candidly offer a single, unassailable, factually accurate conclusion: "We do not know."

Thus Berman agrees with Michio Kaku who says that the question of intelligent design is "undecidable." However, I would want to clarify his assertion that Eastern mysticism teaches that the order and creative process is innate in nature if he means to imply that Eastern mysticism is non-religious. Certainly, even in Theravada Buddhist (the so-called atheist kind), there is a belief in a spirit or soul that is separate from the physical body in that this spirit can be reincarnated in many bodily forms. However, he may be right in the idea that the souls of humans and all living things are not responsible for creating the universe and life itself. That information may reside intrinsically in nature.

Because there is much confusion among the terms mysticism, spiritualism and religion, for the purpose of this paper, let us distinguish these terms knowing full well that I may interchange the terms myself. *Mysticism* has been defined as some phenomenon that transcends rational understanding. Mystical forces may be non-spiritual or spiritual, personal or impersonal, but for the most part they are seen as impersonal. *Spiritualism* is the belief in spirit or soul that can be separated from the physical body and is a part of every known religion. However, spiritualism can exist independently of organized religion in the general belief in spiritual reality without sectarian doctrine to go with it. *Religion* is an organized system of beliefs and dogmas that usually involve some kind of organizational structure. The interpretation of Near Death Experiences by various people who experience them demonstrates the difference between spiritualism and religion. A spiritual person might see a bright light at the end of a tunnel and feel bathed in a warm sense of love, and there may be telepathic communication between this person and the light. However, the light does not take the form of a personal deity - the experience remains generic and in the abstract realm. On the other hand, a religious person might see Jesus in the light, or a Buddhist might see Buddha or Lord Yama. Thus, if a person is deeply indoctrinated into a religion, s/he will superimpose his/her religious images onto the light. The light is somewhat like a Rorschach Test in which a person imposes his/her feelings on a vague, undefined ink blot. Thus, in this sense, a physicist or other person can believe in quantum physics or Eastern mysticism and still claim to be an atheist in not believing in a personal god. An impersonal god is an oxymoron, but what they believe is that there is innate information in nature that creates complex biological forms.

The following table illustrates these differences among mysticism, spiritualism and religion.

Mysticism	Spiritualism	Religion
Belief in mostly impersonal forces in nature	Belief in generic, personal spiritual forces	Spiritual beliefs that have been well-defined as dogma

When humans hit the limits of scientific knowledge, they tend to turn to metaphysics and mysticism. Quantum physics is a good example of this tendency.

CHAPTER 5: CLASSICAL PHYSICS
MY FIRST DEBATE WITH A PHYSICISTS ABOUT LANGUAGE AND LOGIC

Before dealing with modern physics, it is important to discuss classical physics because classical concepts have implications for modern physics. Galileo and Newton's concepts of inertia of rest being equivalent to the inertia of motion played a major role in Einstein's development of relativity, particularly Special Relativity which deals only with inertia of rest and inertial (unaccelerated) motion. My discussion of force and acceleration also relates to Einstein's equating acceleration with gravity.

One force produces straight line motion; two forces are required to produce curved motion.

Centrifugal Force: Fact or fiction?

My first encounter with a physicist outside the classroom was with a gentleman I respect highly. While having lunch with a group of instructors, somehow the issue of *centrifugal force* came up. The physics instructor promptly informed everyone at the table that there was no such thing as centrifugal force – that the perception of this force is an illusion. He went on to say that the only force responsible for circular or curved motion is centripetal force. Now, I had been taught in high school physics that curved motion requires two forces: centrifugal and centripetal. My physics teacher had demonstrated the two forces by swinging a bucket of water over his head. He explained that without centrifugal force, the water in the bucket would fall on his head and that if the centrifugal force was eliminated, he would experience a baptism and a very wet head.

Since I perceived a difference of opinion in the physics community, I felt that there was room for a layman's opinion on this issue. I protested to my physicist friend that with only one force, you get straight line motion, and an object, instead of going into an orbit around another, would simply go straight to the center if there were only centripetal force and no other force to balance it. Besides, didn't Newton say: "A body in motion will continue in motion in a **straight line** unless acted upon by an outside force?" Thus began a dialogue as follows:

Physics guy: Yes, but that other force is centripetal force. The body in motion traveling at a constant rate is not exerting a force. Only accelerated bodies exert force. Centripetal force is an accelerating force, and Newton defined force as F = MA (Mass x Acceleration).

Me: Do you mean to tell me that if I am traveling in my car at a constant speed, say 70 mph, and I collide with a guard rail, that I am not exerting a force on that guard rail and that the guard rail is not exerting a force on my car even though the guard rail is stationary in this frame of reference?

Physics guy: Only accelerated motion produces force – uniform speed is inertia, not force - that's what Newton's formula (F=ma) says.

Me: Let's first define "force" in words, not mathematical formulas, because language (qualitative language that is) must precede math (quantitative language). A force is anything that can change the acceleration or direction of something else – is that a reasonable definition of force?

Philosophy of Science　　　　　　　　　　**CHAPTER 5: CLASSICAL PHYSICS**

Physics guy: Yes.

Me: Then, according to that definition, my car, even though it is traveling at an even speed of 70 mph, is able to change the acceleration or direction of another car that it may, God forbid, collide with. And, a stone wall which my car may collide with can change my car's acceleration (or deceleration) and direction of motion. So, the stationary wall (stationary in this particular box of space) is held together by an electro-magnetic force which resists motion of other bodies that collide with it and can change the acceleration and direction of such bodies. So, according to the verbal definition, a force is the capability of changing the acceleration of something else whether or not it is accelerating. Thus a stationary object, a body moving at a constant speed, and an accelerating body all have force, right?

Physics guy: Still there is no net force without acceleration – you are talking about inertia which is not the same as force.

Me: But if you twirl an object around you with a string, the object on the end of the string has to accelerate first to get into a constant velocity, and once it gets into a constant velocity, it is still accelerating because, by definition, anything moving in a curved line is accelerating, so we have two kinds of acceleration, **straight line acceleration** of an object which is increasing or decreasing in speed and **curved acceleration** where an object is following a curved path (constantly changing direction) and it can be increasing in speed or following a constant speed.

I found that what my physicist friend was teaching his students was pretty much standard in modern physics textbooks, although in older text books, centrifugal force was clearly taught, and I have found "centrifugal force" turning up in the writings of several modern physicists. Here's what *The Physics Classroom* website has to say about the forbidden F-word:

The use of or at least the familiarity with this word centrifugal, combined with the common sensation of an outward lean when experiencing circular motion, often creates or reinforces a common student misconception. The common misconception, believed by many physics students, is the notion that objects in circular motion are experiencing an outward force. After all, a well-meaning student may think, "I can recall vividly the sensation of being thrown outward away from the center of the circle on that roller coaster ride. Therefore, circular motion must be characterized by an outward force." This misconception is often fervently adhered to despite the clear presentation by a textbook or teacher of an inward force requirement. As discussed previously in Lesson 1, the motion of an object in a circle requires that there be an inward net force - the centripetal force requirement. There is an inward-directed acceleration that demands an inward force (1996-2016: paragraph 1).

In the above lesson on physics, the author is clinging to the party-line in adhering to a very narrow definition of force and ignoring Newton's Third Law which is that "For every action there is an opposite and equal reaction." Hence if there is an inward force, there has to be an outward force to balance it and keep a rope (or other flexible line) taut while twirling an object. If there were only an inward force, the object would go straight to the center. When one twirls an object with a rope, the inward pull is balanced by an outward pull like a tug of war between competing teams. If one increases the speed of twirling the outward tension on the rope increases which is balanced by the inward pull of centripetal force – hence for every outward action on the rope, there is an equivalent inward action or force so that the net force is zero and equilibrium is established. However, net zero doesn't mean that there is no force on the rope and the dueling forces can be measured with a scale on each end of the rope. Again, the faster one twirls the scale on the end of the rope, the more force it will register.

In a later conversation with my physicist friend, I added this argument. Now, let's get back to circular motion and centrifugal force. Actually, Galileo described the second force as **tangential force** which operates at right angles to the centripetal force in perfect circular motion. Perhaps the angle would be something other than a right angle with elliptical motion. By definition, the circulating body, representing tangential force, is accelerating because it is constantly changing direction, so we can establish that there is a force there. Of course, the rotating object may be speeding up or slowing down as well as changing directions producing additional acceleration. In any case there is a tangential force operating on the circulating body. This tangential force is opposed by a centripetal force or force toward the center. If the centripetal force is released at a right angle, then the released object travels along a straight line that corresponds to the tangential force. Such was the case with little David when he twirled his sling several times and released the centripetal force holding the stone and hurled it toward Goliath's forehead. We know this is the way slings work whether you believe that story or not.

What happened to the centrifugal force in this scenario? We know that forces and tensions can be distributed in the different directions indicated by vectors. We feel that outward pull when we ride the tilt-a-whirl or merry-go-round at the Fair. This is not an illusion. Again, if you twirl an object on a rope with a scale in your hand, you can measure the outward pull on the rope which has become taut because of this outward force opposing the inward force. If you put a scale on both ends of the rope and twirl one scale around the other, you will see that the force measured in pounds or grams is equal on the two scales. This effect is described in Newton's Third Law of Motion: "For every action, there is an opposite and equal reaction." Put another way the inward (centripetal) force is equal to the outward (centrifugal force) in order to maintain the equilibrium of the circulating body. That means that there are three forces that make up circular or curved motion: centripetal, centrifugal and tangential, and all three of these are vectors since all three represent a magnitude in a specific direction. The centrifugal force is a **derived force** from the interaction of the tangential and the centripetal. As the centripetal force restrains the tangential force of a circulating body, the resistance of the tangential force against the centripetal force distributes some of the force in an outward direction, thus obeying Newton's Third law. The centrifugal force is that reaction that keeps a rope taut when twirling an object on the end of it. Centrifugal force is the reason that centrifuges work which is used to separate heavier substances from lighter ones – such as separating U235 from U238.

Let's see how the direction of these forces are manifested when released from each other. If a pitcher is throwing a baseball, the tangential force causes the ball ideally to head toward home plate because the pitcher is releasing the ball from his hand at a right angle to the centripetal force that holds his arm to his body. The same is true of little David's sling because the sling releases the centripetal force at a right angle to the tangential force. On the other hand, if the orbiting object is released 180-degrees from the centripetal force, the orbiting object will fly outward. If you had a toy train set when you were a kid, you have noticed this effect. When you accelerated the toy train too fast to make the curve, the train would roll outward because as the centripetal force (provided by the track) was disconnected in a direction opposite (180-degrees) from the center of motion (i.e. the center of the virtual circle created by the curve in the track). Of course, if the tangential and derived centrifugal forces wind down and go to zero, an orbiting body will fall toward the center as with an orbiting satellite that falls back to earth when gravity (centripetal force) becomes stronger than the tangential force or angular momentum.

I believe the fallacy about force and motion arises from Newton's indication that there is no net force **ON** a stationary object, and there is no net force **ON** an object in constant motion. A person will not feel a force when a car is sitting still or moving at a constant speed. That does not mean, however, that these inertial objects cannot exert force on an outside object with which there is a collision. We have seen that a stationary object (in a given frame of reference) or an object moving at a constant speed can indeed exert a force on another object with which there is a collision. Recall that a car moving at a constant speed can change the acceleration and direction of a stationary object or another moving object.

Galileo, a predecessor of Newton, had already established the principles of inertia. First of all, Galileo realized that there was no pure inertia of rest. He knew that the earth was spinning and the moving around the sun. However, since that motion is a common denominator to all things terrestrial, we can establish a frame of reference in a relatively closed system where some things are stationary. Since we are riding on the same earth, if we are interacting at the same latitude, then all these other motions are common denominators. With that caveat in mind, he contended that the inertia of rest is equivalent to the inertia of motion because in constant motion there is no internal indication of movement to the system. For example, a cannon ball dropped from the top of a ship's mast will fall straight down whether the ship is anchored at rest or is in constant motion.

However, what the brilliant Galileo failed to take into account was that the cannon ball is attached to the ship and therefore, as part of the ship's system, is carrying the same momentum (force) as the ship; therefore the cannon ball conserves that momentous force and falls straight down as if the ship were sitting still. However, if the moving ship passed under a cannon ball suspended from a stationary crane, and that cannon ball was dropped just above the mast, it would land somewhere behind the mast depending on the velocity of the ship. On the other hand, if the cannon ball were dropped by a stationary crane just above the mast of a ship <u>at rest</u>, the ball would fall straight down by the mast just as if it were dropped from the top of the mast itself. Therefore, from an external reference point, it can be seen that a **ship in a state of inertial rest is different from a ship in inertial motion.** Furthermore, any object in motion is carrying the force that was imparted to it by the force that accelerated it into motion to start with. Thus inertial motion and momentum are equivalent in that they involve constant motion and they exert force in the direction of motion.

Furthermore, if the stationary ship is sitting on a rock, there is no force manifested because the forces of gravity, buoyancy and electromagnetism are in equilibrium, but if the constantly moving ship collides with the same rock, we can clearly see that the moving ship is exerting a force in the direction of movement – a force that has a magnitude (which can be measured) and a direction – thus a vector quantity.

Be that as it may, I have found a very popular physics book that agrees with my analysis that circular motion requires two forces (tangential and centripetal) and not one (centripetal).

The force that accelerates a (circulating) object can be resolved into a component that is **perpendicular** *to the path, and one that is* **tangential** *to the path. This yields both the tangential force, which accelerates the object by either slowing it down or speeding it up, and the radial (centripetal) force, which changes its direction (Sears, Young & Zemansky 1999: pp.18–38).*

Although this source acknowledges the existence of tangential force, other sources still deny the existence of centrifugal force and called it a "fictitious force." I would agree that centrifugal force is a **derived force**, i.e., derived from the tension between centripetal and tangential forces. The feeling of being pushed outward on a merry-go-round is not a delusion. What you are feeling is a tug of war between two forces – you are being torn between a centripetal force that is trying to reel you in and a tangential force that is trying to make you go straight. When the centripetal force tries to pull you in, there is an equal force preventing you from being reeled in, so that some of the tangential force is transferred outward in a centrifugal direction *a la* Newton who said that for every action there is an equal and opposite reaction. This outward distribution of the force is seen in a train wreck or car that doesn't make a curve and tumbles outward. You are literally caught in a tug of war between two forces. Every kid who has played tug of war with a rope knows what happens when you suddenly release the rope – the person on the other end of the rope falls straight back – no mystery there. However, if you are in the center and spinning a kid around with a rope and the kid is leaning back to resist the centripetal pull of the rope, he will fall directly backwards when you release the rope because the centripetal force is released at a 180-degree angle (opposite you).

We quite often hear of the conservation of angular momentum in reference to circular motion as if momentum is not an expression of force. Remember that unless we are in a vacuum, Newton's law that a body in motion will continue in motion unless acted on by a force does not appear to be true. Of course, it is true because we do not live in a vacuum, and a body in motion will stop at some point because that force which acts upon it is friction courtesy of the electromagnetic force. Thus for a body to be in motion on the earth, it had to be acted on by a force, and when it remains in motion till friction stops it, it is carrying or conserving that force which gave it motion to begin with. When a ballerina, twirling with extended arms, pulls her arms into her torso and speeds up, no new force is generated, but the force from her arms which is twirling faster is transferred to the trunk which causes the trunk to spin faster. Thus, the conservation of angular momentum is a *conservation of force* which was initially generated by the ballerina's muscular forces twirling her body with arms extended. The conservation and transference of force is also seen in the Coriolis Effect where winds from equatorial regions migrating to higher and lower latitudes carry their speed to the slower non-equatorial latitudes and exert force on air masses there to speed them up. The Coriolis Effect is indeed force because it changes the direction and the speed of air it impacts. This is true because equatorial latitudes generate more force than higher and lower latitudes by virtue of F=ma and the earth is always accelerating by definition because it is spinning.

The bulge of the earth at the equator also demonstrates centrifugal force. The equator is the

Philosophy of Science **CHAPTER 5: CLASSICAL PHYSICS**

fastest moving part of the earth -- it is rotating faster than latitudes north and south of it because it has to travel farther than other latitudes each time the earth rotates. As a result, the equatorial mass is generating greater force than the mass at any other latitude. This is true because Force = mass x acceleration. Since curved motion is defined as acceleration because the spinning object is constantly changing direction, it fits the formula even though the earth rotates at a fairly constant speed. If the earth's spin were to speed up, then there would be double acceleration. In reality, the earth's spin is said to be slowing down slightly. At any rate, the greater spinning force at the equator acts like a centrifuge in causing the bulge in the belly of the earth – again showing that centrifugal or outward force is a reality, not a pseudo force. Several physicists that I have read use the term centrifugal force, so perhaps they haven't got the memo that there is no such thing. Here's what Howard Hayden, former Professor Emeritus of Physics at University of Connecticut, said. "And the rotating earth is non-inertial, by definition, because centrifugal forces and curvilinear paths are always present (Bethel 2009: p. xix).

In summary, any mass has force associated with it (after all, mass is interchangeable with energy *a la* Einstein).

A stationary mass (in an inertial frame of reference) has force, i.e., electro-magnetic force that holds it together, gravitational force – not to mention the strong force that hold its atoms together and the weak force that tends to make its atoms fission to smaller atoms. It therefore takes force to move or accelerate a still object, and it can definitely change the acceleration and direction of other masses if they collide with the still object. MASS has inherent force.

1) An object or mass in constant motion exerts force in the direction of movement even though there is no force impinging **ON** the object. This can be clearly demonstrated when the constantly moving object collides with another mass.
2) An object accelerating not only has force impinging **ON** it, it also exerts force in the direction of movement.
3) Circular or curved motion requires at least two forces: tangential and centripetal. If there were only centripetal force, the orbiting object would fall directly toward the center of motion. In my opinion, centrifugal force is a force derived from the interaction of centripetal and tangential forces. It is the resistive force that corresponds to Newton's Third Law that balances centripetal force and can be measured with a scale placed at the center and by the amount of tension on a flexible cord attached to the circulating object.

Momentum means carrying force that was imparted to an object. Hence, it is the same thing as inertial motion (uniform speed). Momentum means the conservation of force. For example, the Coriolis Effect is a force that is transferred from the spin of the earth in equatorial regions to higher or lower latitudes. Things do not start in motion without some force being applied to them to accelerate them. When a force is applied to a cue ball, it is transferred to the cue ball by the cue stick which, in turn, transfers the force to the billiard balls.

> **What is force?** Force is a cause, i.e., anything that can produce an effect in something else. A force is what causes acceleration in an object; it does not have to be accelerating itself. For example, gravity is a force that can cause acceleration in an object, but gravity is not accelerating per se. A stationary object (in a given context) can cause an object in inertial

> motion to negatively accelerate if that object collides with it. A still object resists acceleration because it is held together by strong nuclear forces and electromagnetism. Thus an accelerating object is carrying a force, and an object at rest (inertia) is also exerting a force because it resists acceleration and can cause another objects acceleration to change..

Centripetal, Centrifugal and Tangential Forces Demonstrated (see graphic below)

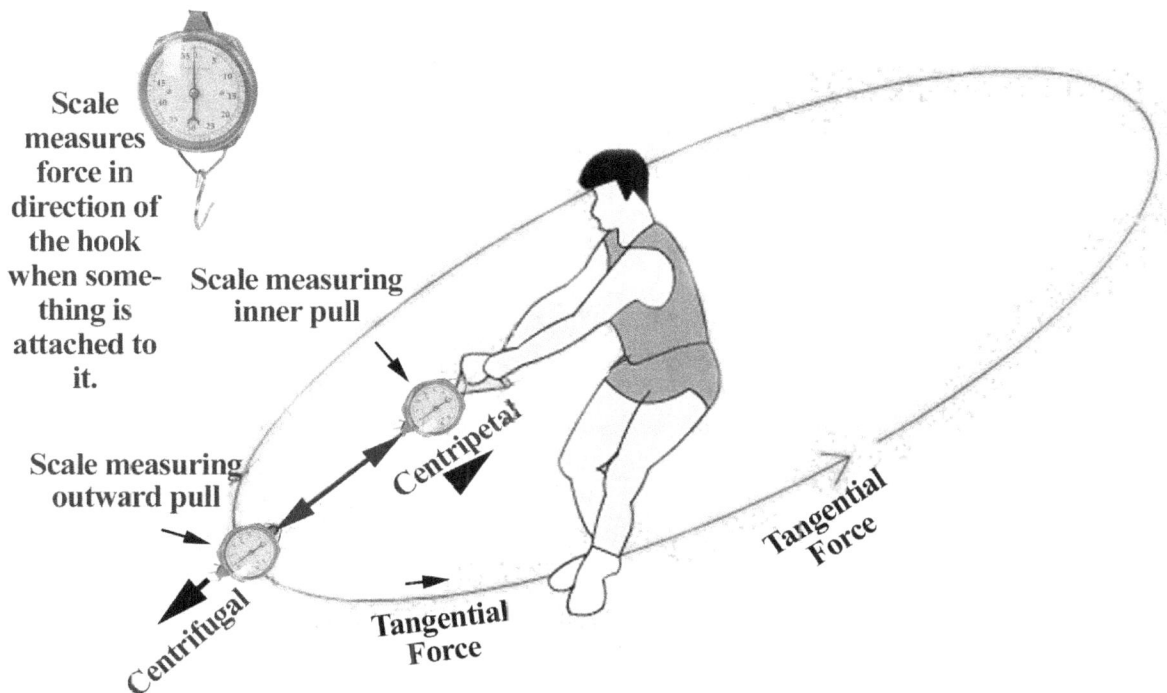

Curved Motion: Centripetal, Centrifugal and Tangential Forces

Scales placed on both ends of cable

To demonstrate the interaction of three vectoral forces that make up curved motion, imagine a hammer thrower, but instead of a ball at the end of the cable, connect a scale that will measure inward (centripetal) force. Also place the same kind of scale in the hands of the hammer thrower to measure the outward pull (centrifugal force). In accordance with Newton's Third Law, the centrifugal pull should equal the centripetal pull as measured on the scales. The tangential force operates at right angles to the centripetal force. The centrifugal force is sometimes called a "fictitious force" and some physicists say it is an illusion. I would call it a "derived" force, and it occurs when the centripetal force opposes the tangential force. A la Newton, when the hammer thrower pulls in against the tangential force, there is an equal pull in the opposite direction to maintain the tautness of the cable. The amount of force measured by the scale will increase as the hammer thrower accelerates the scale, but would even out if the speed of twirling were to become uniform.

Actually each scale measures the tension between centrifugal and centripetal forces since both forces act on each scale as it does on the cable. However, the outer scale is oriented toward the inward tension, and the inner scale is oriented toward the outward tension. As the thrower pulls in on the cable, the scale at the other end is pulling outward, so the scale is measuring combined forces. And, when the scale at the end pulls outward, it is countered by the force pulling inward, so it, too, measures the combined force. As demonstrated, it takes two forces to create curved motion. With one force, you get straight-line motion, so that with centrifugal force only, the ball would go straight to the center. In other words, the hammer thrower would just pull the ball straight to himself if he is exerting only centripetal force and it wouldn't be centripetal at all. He does pull in on the scale, and as he swings the hammer in to his right or left at a right angle to his cable and by holding the cable, he is providing centripetal force. For those who say the hammer (in this case the scale) is not exerting a force, it is accelerating as it makes the circle at right angles to the centripetal force. It is accelerating by constantly changing directions and it is accelerating as the hammer thrower increases the speed of the ball to get the maximum distance. So even if you define force narrowly as F=MA, you have at least two forces: tangential and centripetal. Centrifugal force is derived from the opposition to the centripetal.

Inertia, Momentum, and Acceleration – three forms of force

In my interaction with physicists, my views of physics have been countered by such arguments as there is no force acting upon an object; therefore, the object has no force, and inertia is not force – only accelerating bodies have force because Newton said that Force = mass x acceleration. Of course, the idea of no force acting on an object is false because gravity is acting upon every object in the known universe. In discussing force, one must take into account that all mass has force inherent in it, an object may have force imparted to it and that object will carry that force after the force is no longer applied (i.e., it has momentum), and the mass may be accelerating itself due to internal forces. Let's take examples of each of these.
1) Mass at rest (inertia) has force: Because of electromagnetic bonds and the binding energy of nuclei, any mass is held together by force; therefore, any mass will resist force. A resistance to force is just as much a force as an external source being applied to an object. Resistance to force or acceleration is inertia which is an inherent property of mass. In addition, every mass has some gravity associated with it and is being acted on by the gravity of other objects. If you try to move a big rock that is stationary in earth frame, you encounter resistance equal to its weight. The resistive forces come from internal electromagnetism and the gravity of the earth pulling on the object. In addition, you have friction which is an expression of electromagnetism. So, the big rock has internal and external forces that make it difficult to move.
2) Mass in constant motion (inertia) has force: Some physics texts call this inertia and deny that it is an expression of force. Empiricism must answer the question here. If an object in constant motion collides with a stationary object, we can certainly see that a force was applied to the object because it may have its acceleration changed. An external force may have been applied to the object to set it in motion, but the object is carrying that force imparted to it. Inertia of motion is very similar to momentum (mass x velocity) in that momentum means carrying force that was originally imparted to the object. But something that has momentum may be losing velocity (decelerating) because of resistance such as friction. In that case, we say the object is losing momentum. Herein lies the difference between inertia of motion and momentum. Inertia of motion would mean that the velocity is constant; whereas in momentum, the object may lose velocity and therefore not be constant. Actually, even a constant rate of motion should not be

referred to as inertia. This type of motion should be referred to as momentum because the object is carrying the motion that was given to it by acceleration at an earlier time. If all motion began with the Big Bang, then all motion began as acceleration. If a ship is accelerated in the vacuum of space (where there is no resistance) and the force that is accelerating it is removed, then the ship will continue at the last rate of speed imparted to it. but not accelerate further.

3) Mass that is accelerating is carrying force (F-=MA). Some physicists say that this is the only expression of force, but I have given examples above to show that this is not true. Since deceleration is called negative acceleration, it is ironic that an object that is decelerating would exert a force; whereas an object in constant motion would not according to Physics doctrine. So, if I hit a baseball and as it is decelerating it hits you, it is exerting a force on you. If, on the other hand, an object in constant motion hits you, it shouldn't hurt you because it is not exerting a force.

Forms of Force

	Mass at rest (Inertia of rest)	Mass in constant motion (Inertial motion)	Mass in Motion Momentum	Mass accelerating
Resists Force	Yes	Yes	Yes	Yes
Carries Force		Yes	Yes	Yes
Exerts Force	Yes	Yes	Yes	Yes

So, from the above table we see that mass or matter can exert force at any speed or no speed at all in a given frame of reference.

The following graphic will demonstrate that inertia of rest, inertial motion and accelerated motion are each a type of force. It also demonstrates that inertia of rest is not the same thing as inertia of motion.

How much force would it take to push a 3-ton pickup backward 10 feet in 5 seconds in these three scenarios

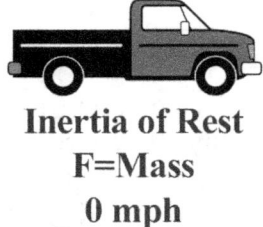

Inertia of Rest
F=Mass
0 mph

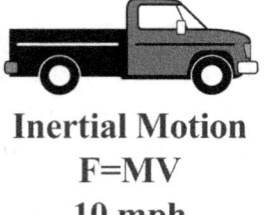

Inertial Motion
F=MV
10 mph

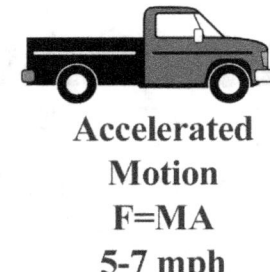

Accelerated Motion
F=MA
5-7 mph

Here we see that in each situation, it requires force to move the pickup; therefore anything that resists force is a force in itself whether the object is at inertial rest, in inertial motion or in accelerated motion. This illustration also shows that inertia of rest is not the same thing as inertial motion.

---

*Inertia resists force and force resists inertia. Old riddle: What happens when an irresistible force meets an immovable object (infinite inertia)?

*Another piece of evidence that stationary masses exert force is Einstein's famous formula: $E=MC^2$. Assuming that energy is a synonym for force, Einstein is saying that energy and force are built into the mass. This force is in the form of electro-magnetism and the strong force that holds atomic nuclei together. Therefore, when something collides with a stationary object, the object resists the motion of the moving object and can change its direction and acceleration. Thus, a stationary object is constantly exerting force because of the energy inherent in it.

To sum up:
1>Inertia should refer only to bodies at rest in a given context therefore, Inertial Force = M (for mass).
2>Momentum should refer bodies moving at a constant speed because they are carrying the force imparted to them by acceleration: Therefore: p (momentous) force = mv
3>Acceleration is described by Force = ma.

Thus we can say that inertia resists acceleration and acceleration resists inertia. When motion is given to an object by an accelerating force, the object continues to carry that force even though its velocity may even out. This continuation of force and velocity is momentum.

Angular Momentum and Twirling Ballerinas

Once I was in a group where a physicist explained the dynamics of what makes a ballerina twirl faster when her arms are pulled in. The physicists' explanation was that there was no force involved in this increase in speed, but that the effect was caused by the conservation of angular momentum (indicating that momentum is not force). Of course, there is no contradiction between momentum and force since momentum involves an object continuing to carry a force through time, so momentum is an expression of force just as inertia and acceleration are expressions of force. Since curved motion is defined as *acceleration* whether speeding up, slowing down or moving at a constant rate, then a twirling ballerina is exerting force since she is a mass moving in a circle, and force is equal mass times acceleration mathematically defined. To get her started from a still position requires a force which accelerates her, and to continue twirling requires that the force continues to operate. This carrying of the force forward in time is what should be termed momentum. The motion of the ballerina can be explained in terms of centripetal and tangential forces that are always present in curved motion around a center – in this case the center of gravity. The following are the elements involved in the dynamics of twirling or spinning.

1>A tangential force is generated when the ballerina extends her arms and one leg outward.
2>That force is transferred to curved motion by twisting her body and by the fact that the forearm travels in an arc from the elbow outward.
3>When she draws her arms and leg inward near the center of gravity, the greater speed on the periphery is transferred to the trunk of the body thus increasing her speed of spinning. Although her spin accelerates, there is no new force in the whole system. She is simply transferring the greater force from the periphery closer to the trunk and the center of spin.
4>As indicated, the tangential force is created when the ballerina slings her arms and leg outward; the centripetal force that opposes the tangential force is caused by the arm and leg's attachment to the trunk of the body which keeps the tangential force constrained in circular motion.
5>What stops the spin of the ballerina is friction of the floor against the foot and the re-extension of the arms and the leg to transfer the force from the trunk to the periphery, thus slowing the spin of the body.

The body of the ballerina can be considered one system. Thus, in the process of increasing the speed of spin, no new force is generated in the system; it is simply transferred from the periphery closer to the center where it increases the speed of the center. This phenomenon involves the same principle as the Coriolis force which some call an *effect* and insist that it is not a force. Again, getting the semantics straight is the key to understanding this phenomenon and force is involved in this effect.

Coriolis Effect or Force?

When winds traveling with the spin of the earth at the equator move northward or southward away from the equator, they become like eddies or a whirlpool. In the Northern Hemisphere, the winds spin counterclockwise (from a satellite perspective above the earth) and clockwise in the Southern Hemisphere. Since the winds are carrying the momentum of the spinning earth which holds the atmosphere with little slippage, they tend to move from West to East or more properly, from West to Northeast in the Northern Hemisphere and from West to Southeast in the Southern Hemisphere. But what causes the atmosphere to deviate from the equator to begin with? The probable answer is that there is a lower pressure to the North or South and atmospheric molecules seek equilibrium (or entropy). Another possible factor is that there is some slippage between the solid/liquid earth and the atmosphere which generates differential speeds between the lower levels of atmosphere which is more affected by gravity and the higher layers of atmosphere which are subject to lesser pull toward the center of the earth.

Thus, a hurricane formed in the Northern Hemisphere can track westward instead of eastward resisting the force of gravity and the momentum imparted by the spinning earth. If the atmospheric density or pressure is less to the West, it will track in that direction thus resisting momentum. When the winds from the equator migrate to a higher or lower latitude, they appear to increase in speed because latitudes on both sides of the equator are spinning slower than the equator because their spin describes a smaller circle with each rotation of the earth. So, from the point of view of a higher or lower latitude, there is an increase in speed relative to the ground and therefore a greater relative force. However, considering the earth as one system, there is no increase in speed or force. Like the ballerina who pulls her hands and arms from a greater circle to a smaller circle, there is an increase in speed, but no new force is generated considering her whole body as a system. However, the Coriolis Effect does involve force because the force of moving winds is carried from one latitude to another. Also, since the winds are spinning, that is a form of acceleration and the air molecules do have mass, so again force=mass x acceleration. Certainly, we would not want to deny that hurricane winds are carrying force. Of course, that force is imparted not only by gravity and the spin of the earth, but by temperature-humidity differentials which create high and low pressures.

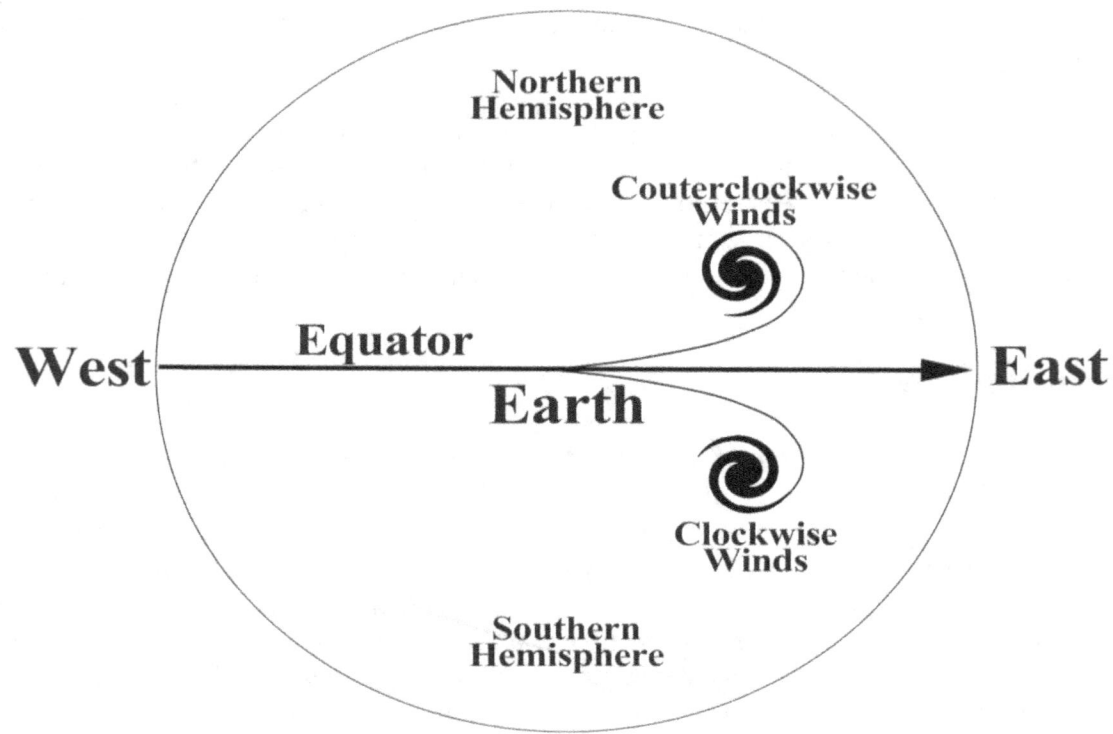

Philosophy of Science   CHAPTER 5: CLASSICAL PHYSICS

Inertia of Rest Distinguished from Inertial Motion
As indicated above, inertia of rest is similar to inertia of motion, but they should not be seen as equivalent. Considering them to be equivalent leads to a number of problems in physics theory. Consider this graphic also:

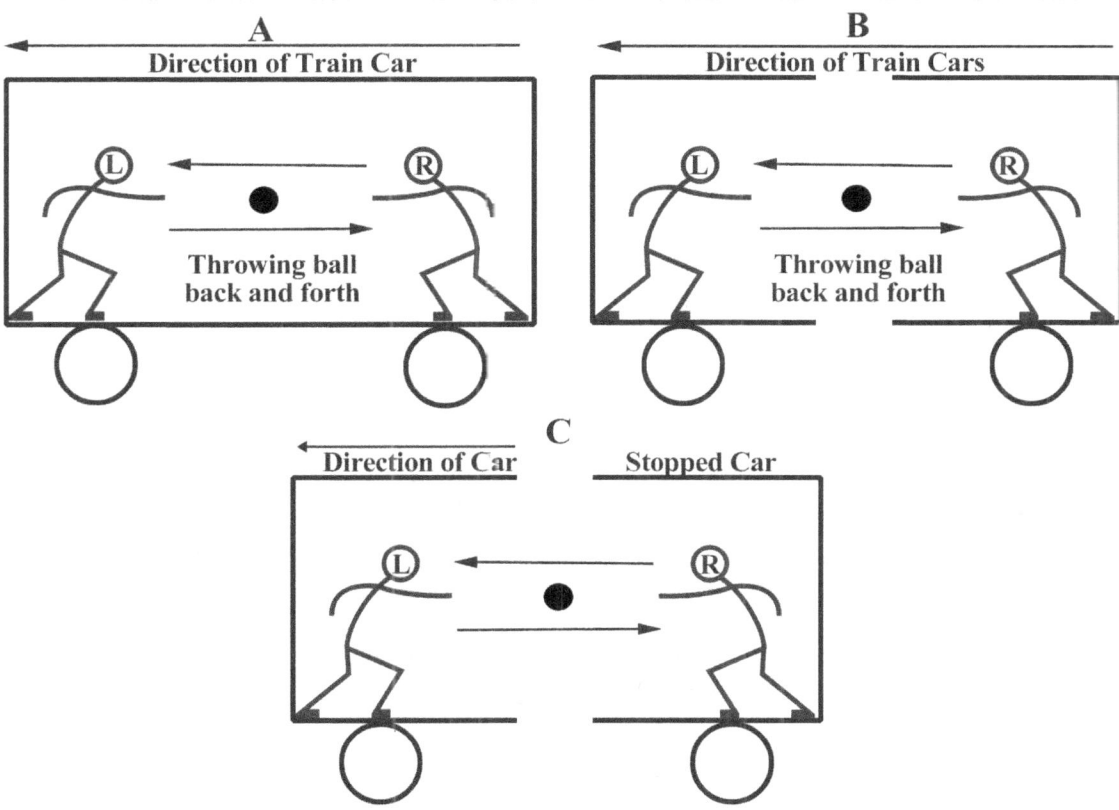

In the illustration above, we see Leftie (L) and Rightie (R) throwing a ball back and forth on a train car and then between train cars.

In scenario A, Leftie and Rightie are in the same box car throwing the ball back and forth to each other at the same speed. The speed of the ball does not seem to be affected by the speed of the car because the car's speed is inertial (unaccelerated). When R throws the ball to L in the direction the car is moving, the ball speeds up because it is carrying the momentum of the car, but L is also carrying the momentum of the car and is thus moving away from the ball at the same rate as the ball's speed has been increased by the car. Therefore, the added momentum of the ball is equivalent to the added momentum of L and therefore the result is the same as if they were both on a stationary car. **Conversely,** when L throws the ball to R, the speed of the car subtracts from the speed of the car since the call is traveling against the motion of the car. However, R is moving toward the ball at the same rate that the ball is being slowed by the motion of the train. Again, equivalence occurs and the result is the same as if the car were stationary.

Philosophy of Science CHAPTER 5: CLASSICAL PHYSICS

In Scenario B, the same dynamics of inertial would occur if L and R were passing the ball between two cars traveling at the same constant rate of speed (discounting any aerodynamics caused by the moving cars through the atmosphere.

In Scenario C, R's car suddenly stops and is at rest. Now we see that the ball does not behave in the same way between a car in inertial motion and a car at inertial rest. The ball would be slowed in its speed when thrown from L toward R, and if it reached R, it would be slower than in scenarios A and B. Therefore, we can see that inertial motion is not the same thing as inertial rest. Also if the moving car collides with a stationary car on the tracks, force will be applied whereas if a car is placed in a still position in contact with another stationary car, no force will be manifested since all forces are balanced.

Now, the relativist would say that there is no privileged perspective and the view inside the box car is equally as valid as the view outside the train. I would disagree with this statement and say that the more comprehensive the point of view, the better it is. The sighted person who can see a whole elephant has a perspective superior to a group of blind men who are touching different parts of the elephant. If a box of space is specified as a context (as Einstein does in his thought experiments), then the outside-the-train observer can see the compound motion of the ball's speed combined with the speed of the train whereas the people inside an enclosed box car can see only the motion of the ball. So, in a given context, space can be seen as absolute and a grid can be constructed to relate space and time. When Einstein says that there is an observer at rest and one in inertial motion, he is, in fact, admitting that within that context, space is treated as absolute – otherwise it would not be possible to determine who is at rest and who is in motion. Since the at-rest observer and the observer in motion cannot determine who is moving (if they are blind to the background environment), the outside observer can detect motion (and Einstein in this thought experiment sees himself as that outside observer who can determine that one observer is at rest and the other in motion).

The relevance of classical theory will become apparent as we begin to discuss the concepts of force, inertia and acceleration in relativity theory. Surely, if physics can be this confounded about something as basic as circular motion, there is room for an outsider's opinion with a fresh perspective on this and other theories.

Philosophy of Classical Physics vs. Philosophy of Modern physics

While some physicists deny the notion that classical physics uses a different kind of logic and philosophy than modern physics (relativity and quantum physics), it seems clear to this author that there is a difference between classical science based on rational-empiricism or logical-positivism and modern physics. Classical science is based on the notion that there is a mind-independent objective reality and that theories must be tested against this reality. In short, there must be physical or material evidence for a theory if it is to meet the test of science. Thus, science is, indeed, based in the philosophy of materialism. Even parapsychology, which some physicists view as pseudo-science, requires physical evidence for so-called non-physical phenomena – the psi factor. On the other hand, the Copenhagen interpretation of quantum physics (at lease the Heisenberg version) indicates that the mind is part of the quantum process (observation is needed to collapse a probability wave). In maintaining this subjectivism,

Heisenberg is indicating that, at least at the quantum level, there is no mind-independent, objective reality.

Contrastingly, Neils Bohr saw quantum mechanics as a generalization of classical physics although it violates some of the basic ontological principles on which classical physics rests. It is interesting that Bohr rejected the subjectivism of Heisenberg, but he was not a logical-positivist either (Popper 1967). Some of these principles of classical physics that different from quantum physics are identified by Jan Faye (2014):

The principles of physical objects and their identity in classical physics:
1~Physical objects (systems of objects) exist in space and time and physical processes take place in space and time, i.e., it is a fundamental feature of all changes and movements of physical objects (systems of objects) that they happen on a background of space and time;
2~Physical objects (systems) are localizable, i.e., they do not exist everywhere in space and time; rather, they are confined to definite places and times;
3~A particular place can only be occupied by one object of the same kind at a time;
4~Two physical objects of the same kind exist separately; i.e., two objects that belong to the same kind cannot have identical location at an identical time and must therefore be separated in space and time;
5~Physical objects are countable, i.e., two alluded objects of the same kind count numerically as one if both share identical location at a time and counts numerically as two if they occupy different locations at a time;
6~The principle of separated properties, i.e., two objects (systems) separated in space and time have each independent inherent states or properties;
7~The principle of value determinateness, i.e., all inherent states or properties have a specific value or magnitude independent of the value or magnitude of other properties;
8~The principle of causality, i.e., every event, every change of a system, has a cause;
9~The principle of determination, i.e., every later state of a system is uniquely determined by any earlier state;
10~The principle of continuity, i.e., all processes exhibiting a difference between the initial and the final state have to go through every possible intervening state; in other words, the evolution of a system is an unbroken path through its state space; and finally
11~The principle of the conservation of energy, i.e., the energy of a closed system can be transformed into various forms but is never gained, lost or destroyed.

Yours truly is in agreement with Faye in identifying these differences and their implications for the philosophy of science and the scientific method. Modern physics has made a turn toward pure metaphysics and even mysticism as Bohm and Capra have acknowledged. The mantra frequently heard to harmonize this discrepancy between the classical philosophy of science and modern physics is that the quantum world operates on different principles (or perhaps laws) than the macro-classical world. However, when quantum physics makes a prediction about events in the macro world (the cat is dead and alive) which is not supported by observation, then the philosopher of science is in a quandary as to which is true - or can it be that both are true. When confronted with the contradictory statement that a quantum is both a particle and a wave at the same time and the wave can spread throughout the universe, we hear at least four different ways to resolve this dilemma.

1~The particle is somewhere in the wave – it has a probability of being more in the crest of the wave that the trough according to some.
2~The particle actually exists as a wave simultaneously.
3~The quantum is a wave until an observation is made and then it condenses into a particle (Copenhagen Interpretation).
4~The wave never collapses and the particle (and the whole universe connected to it) splits so that each probability is actualized in another universe identical in every respect to the other universes except for the alternate probability state (Many Worlds Interpretation).

The fact that there are so many contradictory interpretations of quantum mechanics indicates to this author that the theory is not nearly as solid as quantum physicists claim. This author believes that the picture and functioning of the atom produced by quantum mechanics is solid and accords with macro observations such as chemistry experiments. However, the theories derived from the double-slit experiments do not accord with predictions they make about the macro world and thus violate the classical scientific method of empiricism.

SUMMARY

The thesis of this work throughout has been that language is the basis of theory and the foundation of logic. Thus, a good theory begins with precise language and proceeds to mathematical calculation and empirical testing. The prevailing philosophy of science is based upon the dual pillars of rationalism and empiricism (or logical positivism), and language is thus the primary tool of these two pillars. The principle critique of Modern Physics in this book is that physicists have played fast and loose with language and logic and therefore have taken indecent liberties with the facts. Moreover, some Modern Physicists deny the existence of material world (e.g., Zero Worlds Theory) which is the foundation of empiricism. Science demands physical evidence for any theory proposed, and if one denies the reality of the physical world, s/he has rejected one of the two pillars of the scientific method. Denying the existence of the material world thus leads one into metaphysics and mysticism and a departure from science - since a theory or hypothesis cannot be falsified in the absence of material evidence. More and more theories of Modern Physics, such as the Many Worlds Theory and String Theory, are said to produce no physical predictions and are thus untestable. Therefore, they fail to meet the test of falsifiability and hence are unscientific.

The key terms that are misused in relativity that lead to false conclusions are space and time. Space, properly defined, is the absence of matter and energy. While outer space is said to have about three atoms per cubic meter, there is certainly more space than matter (low density), the farther one gets from stars, galaxies and galactic clusters. It is said that there is a bubble of virtually pure nothingness between galactic clusters. Michelson-Morley's experiment is accepted as strong evidence that space is not ether; Einstein's Special Relativity indicates that space is vacuum; General Relativity indicates that space and spacetime are like physical media that, when curved by mass, force planets to revolve around stars; and Quantum Theory asserts that space is something and nothing at the same time. Quantum theorists tell us that particles are constantly bubbling up out of the vacuum (something from nothing). Two other concepts about space are that space is information and space is structure. These ideas are abstractions and, therefore, do not involve physical media and are thus unscientific since, again, science demands physical evidence for concepts and theories. There must be a physical medium to carry information and to provide structure.

Allow me to sketch out, in the following table, the progression from physics to metaphysics to mysticism, culminating in String theory, the most purely metaphysical theory of all. String Theory was not hatched in a vacuum but was the logical extension of a trend to substitute mathematical proofs, and even mysticism, for empirical evidence. The entries in the table below need much qualification and clarification, which this book has gone to great lengths to provide, but this graphic puts my philosophical critique of physics in the neatest nutshell that I could conjure.

Philosophy of Science   SUMMARY

	Classical Physics	Special Relativity	General Relativity	Quantum Mechanics	String Theory	Cosmology
Physics (rational-empiricism)	Theory tested via Experiments Observations	Thought Experiments & some Experiments Observations ~Length Contraction untested	Thought Experiments & some Experiments Observations	Mathematical Modeling & Some Theory Testing via Experiments Observations	*No* Theory Testing via Experiments Observations	Some Theory Testing via Observations
Metaphysics (Pure rationalism)	Ether	Space and Time are material realities, but space is vacuum	Spacetime curvature is physical cause	Heavy reliance on mathematical formalism (e.g., MWI)	Pure Mathematics	Singularity in spacetime
Mysticism (Non rational)	Newton's theological interpretations	Twins paradox	Theory of gravity says there is no gravity.	Consciousness determines physical reality & Contradictions allowed in theory (cat dead and alive)	1-D strings & 11 dimensions	Something from nothing

The above table shows the progression (regression) from physics to metaphysics to mysticism as the field of physics moves in time from classical to relativity to quantum mechanics and string theory.

1>Although **Classical Physics** has the metaphysical concept of ether, an imaginary, invisible medium which supposedly carries electromagnetism and gravity, classical theories are decidedly more empirical than Modern Physics beginning with relativity. While one might expect that theories of physics would become more scientific as they evolve through time, they have become less empirical and more unscientific with the flow of time in the considered opinion of this author.

2>In **Special Relativity**, Einstein, as a theorist, relied more on thought experiments than real experiments in formulating his theories. The notion that space and time (abstractions) are real physical entities on the same order as mass and energy is metaphysical thinking that leads to fallacious conclusions. The notion that space and time are material things leads to the fallacy that they are elastic and can be stretched and contorted by the forces of mass and energy to fit one's theory. And, the idea that inertia of rest is the same as inertial of motion leads to the mystical absurdity of the twins paradox – that each twin is younger than the other because the inertially-moving twin can't tell whether he is moving or his stationary twin is moving. The theory of length contraction, necessary to maintain the constancy of spacetime, has never been tested. The most accurate clock in the world (the Cesium clock) is not accurate enough to test the hypothetically small deviations in time cause by the speed of our fastest vehicles. Physicists claim that Cesium clocks are accurate to two nanoseconds per day, but Hafele and Keating indicate that Cesium clocks can deviate from each other as much as one second per day – far too inaccurate to verify their findings on time dilation produced by the speed of airliners. The most glaring contradiction in Special Relativity is that space and time are material entities in one part of the theory, yet space is seen as a vacuum in another part of the theory. The initial premise of the theory is that the speed of light is constant in a vacuum – meaning that space is nothingness, not a material entity.

3> **General relativity**, the merging of space and time into one entity, spacetime, is a metaphysical construct that Einstein rejected initially. The mystical aspect of General Relativity is that the motion of smaller bodies around larger bodies is caused by the grooves in spacetime – not some independent force called gravity. Yet GR is considered the best theory of gravity that physics has to offer. If the attraction between bodies is caused by the curvature of spacetime, then there is no need to search for the graviton, the hypothetical particle of gravitational energy. Thus, unification theories can proceed without the incorporation of gravity into the Standard Model.

3> **Quantum theory** takes a quantum leap into metaphysics and mysticism. Although there is some observation and experimentation, each of the various interpretations with their mathematical formalisms are said to be equally valid even though they logically contradict each other. For example, how can the mathematics of the Copenhagen Interpretation be as valid as the math of the Many Worlds Theory when Copenhagen says that the wave of probabilities collapses to one definite state in this one world, and the Many Worlds says that the wave of probabilities splits the universe into multiverses in which all possibilities are realized in some universe somewhere? Quantum theory wades into mysticism when it holds that consciousness determines the outcome of experiments and that a particle in a state of superposition can be in many places at once leading to the absurd conclusion that a cat can be dead and alive if its life depends on the uncertainty of a particle's state. Thus, contradictory concepts can coexist in the same quantum theory as in mysticism.

4> **String theory**, the ultimate quantum theory, carries the metaphysical and mystical trends in physics to their logical (illogical) extremes. The theory is based on tiny imaginary one-dimensional strings that vibrate in 11 dimensions to produce the particles and energy frequencies of the larger world. The theory is based on pure mathematical metaphysics, producing no testable hypothesis or predictions. It is mystical in the sense that 1-dimensional strings could not exist as matter and could never add up to 3-dimensional matter. The contradiction that matter can be 1-D and 3-D at the same time cannot be logically resolved.

5> **Cosmology** borrows metaphysical and mystical baggage from quantum theory and relativity to explain large-scale, universal phenomena. The singularity in spacetime (a dimensionless point) is borrowed from General Relativity and is obviously a metaphysical construct since matter-energy cannot logically exist in non-space and non-time. The mystical idea of getting something from nothing is borrowed from quantum physicists who inform us that virtual particles can pop into existence out of the vacuum (which has zero-point energy) and can become real particles of matter under certain circumstances. This quantum foam, as it is called, gives a rationale for how the Big Bang could have banged into existence from nothing.

In the final analysis, then, this increasing trend toward metaphysics and mysticism has pushed physics over the precipice to its logical conclusion - physics is math and mysticism. I agree with Fritjof Capra (1999) when he said:

Science does not need mysticism,
and mysticism does not need science,
but man needs both.

However, despite his wise saying, Capra injected mysticism into science and defended its infiltration into Modern Physics. Contrariwise, I think mysticism should be free of science, and *science should be free of mysticism* – the twain should never meet nor be mixed.

References

Abbott, B. (2007). "Microwave (WMAP) All-Sky Survey". Hayden Planetarium.

Aharonov, Yakir and Rohrlich, Daniel (2005) *Quantum Paradoxes: Quantum Theory for the Perplexed.* Wyley VCH

Aharonov, Yakir (2002) *Uncertainty.* Discovery Science Video.

Albert Einstein Site Online. Retrieved from http://www.alberteinsteinsite.com/quotes/einsteinquotes.html.

Alok, Jha, (August 6, 2013). "*One year on from the Higgs boson find, has physics hit the buffers?*". The Guardian: London.

Ananthaswamy, Anil (2017) *A classic quantum test could reveal the limits of the human mind.* New Scientist. https://www.newscientist.com/article/2131874-a-classic-quantum-test-could-reveal-the-limits-of-the-human-mind

Ashby, Neil (2002) "Relativity and the Global Positioning System." *Physics Today*, May 2002, p. 41.

Barish, Barry C. and Weiss, Rainer (October 1999). "LIGO and the Detection of Gravitational Waves". Physics Today. 52 (10).

BBC Documentary (2008) *Hawking Radiation.* Interview with Professor Bernard Carr, Queen Mary University of London. https://www.youtube.com/watch?v=S6srN4idq1E

Beckmann, Petr (1987) *Einstein Plus Two.* Boulder, CO: Golem Press.

Behe, Michael J. (1996). *Darwin's Black Box: The Biochemical Challenge to Evolution.* New York: Free Press.

Berman, Bob (Dec. 2017) "Intelligent Design" in *Astronomy* Magazine.

Bethel, Tom (2009) *Questioning Einstein: Is Relativity Necessary?* Pueblo, CO: Vales Lake Publishing.

Biever, C. (6 July 2012). "It's a boson! But we need to know if it's the Higgs". New Scientist. Retrieved 2013-01-09.

Bishop, Owen (1984) *Yardsticks of the Universe.* New York: Peter Bedrick Books.

Boesgaard, A. M. and Steigman, G. (1985). "Big Bang Nucleasynthesis: Theories and

Observations", *Ann. Rev. Astron. and Astrophys.* 23, 319.

Boswell, J. (1823). The Life of Samuel Johnson, vol. 1. London: J. Richardson & Co.

Bohm, David (1980) *Wholeness and the Implicate Order*. New York: Routledge & Kegan Paul.

Brainy Quotes: http://www.brainyquote.com

Brandenburg, John (2011) Beyond Einstein's Unified Field. Kempton, IL: Adventures Unlimited Press.

Brown, James Cooke (1975) *Loglan 1: A Logical Language,* Loglan Institute.

Bryce, Emma (2016) *Will Wind Turbines Ever be Safe for Birds.* http://www.audubon.org/news/will-wind-turbines-ever-be-safe-birds

Bunge, Mario (2001). *Philosophy in Crisis: The Need for Reconstruction.* Amherst, New York: Prometheus Books.

Capra, Fritjof (1999) *The Tao of Physics*. Boston: Shambhala Publications.

Carrey, Jim (1995) *Ace Ventura: When Nature Calls*. Movie. Morgan Creek Productions.

Carroll, Sean (May 20, 2013) *Arrow of Time - Sixty Symbols*. https://www.youtube.com/watch?v=9VFGuupXwng#t=235.460771

Carroll, Sean (Jan 22, 2013) *Quantum Mechanics (an embarrassment) - Sixty Symbols*. https://www.youtube.com/watch?v=ZacggH9wB7Y

Carroll, Sean M. (2006). *C-SPAN broadcast of Cosmology at Yearly Kos Science Panel, Part 1*.

Carroll, Sean M. (2004). *Spacetime and Geometry*. Addison Wesley.

Case, Thomas. (2013) *A Short Introduction to Metaphysics*. (Kindle Locations 49-50). Didactic Press. Kindle Edition.

Caughill, Patrick (July 28, 2017) *A New Breakthrough in Quantum Computing is Set to Transform Our World.* Futurism website: https://futurism.com/a-new-breakthrough-in-quantum-computing-is-set-to-transform-our-world/

Cherenkov, Pavel A. (1934). "Visible emission of clean liquids by action of γ radiation". Doklady Akademii Nauk SSSR 2: 451. Reprinted in Selected Papers of Soviet Physicists,Usp. Fiz. Nauk 93 (1967) 385. V sbornike: Pavel Alekseyevich Čerenkov: Chelovek i Otkrytie pod redaktsiej A. N. Gorbunova i E. P. Čerenkovoj, M.,"Nauka, 1999, s. 149-153.

CERN Document Server, http://home.cern/topics/antimatter

Cervantes-Cota, J.L.; Galindo-Uribarri, S.; Smoot, G.F. (2016). "A Brief History of Gravitational Waves". *Universe* 2 (3): 22. doi:10.3390/universe2030022

Chomsky, Noam (1957) *Syntactic Structures*. The Hague: Mouton.

Clegg, Brian (2012) *Gravity: How the Weakest Force in the Universe Shaped Our Lives*. St. Martin's Press. Kindle Edition. (pp. 149-150, Kindle location 1787).

Clegg, Brian (2014). *30-Second Quantum Theory*. New York: Metro Books.

Cox, Brian (2011) *The Quantum Universe. (And Why Anything That Can Happen, Does)* Da Capo Press. Kindle Edition.

Cox, Brian and Forshaw, Jeff (2010) *Why Does $E=MC^2$*. Da Capo Press, Kindle Edition.

Davies, Paul (2003) *How to Build a Time Machine*. Penguin Publishing Group. Kindle Edition.

Davies, Paul and Gregersen, Niels Henrik (2010) *Information and the Nature of Reality: Physics to Metaphysics*. New York: Cambridge University Press.

Davies, Paul (2006) Interview in: "The Anthropic Principle" Video by IBBC Worldwide Ltd.

Del Rosso, A. (19 November 2012). "Higgs: The beginning of the exploration". CERN Bulletin. Retrieved 2013-01-09.

Deutsch, David *(1998) The Fabric of Reality: The Science of Parallel Universes. Amazon Kindle Book.*

Deutsch, Sid (2005) *Einstein's Greatest Mistake: The Abandonment of the Ether*. New York: iUniverse, Inc.

Dirac, P.A.M (May 1963) "The Evolution of the Physicist's Picture of Nature," in Scientific American, May 1963, p. 53.

Discovery Science Video (viewed 2002) *Uncertainty*.

Dyson, F.W.; Eddington, A.S.; Davidson, C.R. (1920). "*A Determination of the Deflection of Light by the Sun's Gravitational Field, from Observations Made at the Solar eclipse of May 29, 1919*". Phil. Trans. Roy. Soc. A 220 (571-581): 291–333.

Eagle, Bob, aka Dr. Physics (April 9, 2012) *Bell's Inequality*. https://www.youtube.com/watch?v=qd-tKr0LJTM

Einstein, Albert (1916) *Memorial Notice for Ernst Mach,* Physikalische Zeitschrift 17: 101-02.).

Faye, Jan (Fall 2014 Edition) "Copenhagen Interpretation of Quantum Mechanics", *The Stanford Encyclopedia of Philosophy* Edward N. Zalta (ed.). https://plato.stanford.edu/archives/fall2014/entries/qm-copenhagen

Feynman, Richard P (1990) *QED, The Strange Theory of Light and Matter*, Penguin, p. 128

Feynman, Richard (1965) *The Character of Physical Law.*

Folger, Tim (June, 2002) *Does the Universe Exist if We're Not Looking?* Discover Magazine.

Frank, Adam (2016). *Three Atoms per Cubic Meter*. In NPR Blogger 13.7: Cosmos and Culture. Retrieved from http://www.npr.org/2016/08/09/489361654/short-answers-to-big-questions-exploring-atoms-in-space.

Gardner, Martin (1997) Relativity, Simply Explained. Dover Publishing.

Garret, Ron (2011) *The Quantum Conspiracy: What Popularizers of QM Don't Want You to Know*. Google Tech Talk. https://www.youtube.com/watch?v=dEaecUuEqfc.

Gawiser, E.; Silk, J. (2000). "The cosmic microwave background radiation". *Physics Reports* 333–334: 245–267.

Gibbs, Phillip (1996). "*Can Special Relativity Handle Accelerations?*" The Original Usenet Physics FAQ. Retrieved 2014-07-23.
http://math.ucr.edu/home/baez/physics/Relativity/SR/acceleration.html

Gingerich, Owen (1992) *The Great Copernicus Chase*. Sky Publishing Corp.http://astro.wsu.edu/worthey/astro/html/im-lab/stonehenge/stonehenge.html

Goldsmidt, Walter (1970) *Whatever Happened to Human Nature*. Lecture at Wake Forest University.

Good Reads: http://www.goodreads.com/quotes

Greene, Brian (1999) *The Elegant Universe.* New York: W.W. Norton and Company.

Greene, Brian (1999) *The Elegant Universe: Superstrings, Hidden Dimensions, and the Quest for the Ultimate Theory.* W. W. Norton & Company. Kindle Edition.

Greene, Brian (2016) *M-Theory, String Theory and the Elegant Universe.* Discovery Channel - String theory rare documentary, National geographic retrieved from Youtube: . https://www.youtube.com/watch?v=qtaAM84Kt2I.

Greene, Brian (2004) *The Fabric of the Cosmos.* New York: Random House.

Greene, Brian (2011). *The Fabric of the Cosmos:* Video by NOVA and PBS.

Greene, Brian (2014). *The Fabric of the Cosmos: What is Space?* Youtube: Rising Life Media.

Gribbin, John R. (1984) *In Search of Schrodinger's Cat.* Bantam Books.

Gribbin, John R. (2009) *In Search of the Multiverse.* John Wiley and Sons.

Gules and Sable: *Escutcheons of Science*

Gunion, John F.; Haber, Howard; Kane, Gordon; Dawson, Sally (2008*)* *The Higgs Hunter's Guide.* Westview Press.

Hadhazy, Adam (2016, December*)* *Nothing Really Matters.* Discover Magazine.

Hameroff, Stuart (2008). "That's life! The geometry of π electron resonance clouds". In Abbott, D; Davies, P; Pati, A. Quantum aspects of life (PDF). World Scientific. pp. 403–434.

Hatch, Ronald (1995) *"Relativity and GPS,"* Part I, Galilean Electrodynamics, 6, 3, pp. 51-57, and Part II, Ibid. 6, 4, pp. 73-78)

Hatch, Ronald R (1995) *Relativity and GPS,* Part II, Galilean Electrodynamics 6, 4 , p. 73-78/ http://aetherforce.com/the-suppression-of-inconvenient-facts-in-physics-by-rochus-boerner/

Hatch, Ronald (2004) *"Those Scandalous Clocks."* GPS Solutions: 67-73, p. 72.

Hayden, Howard and Whitney, Cynthia (1990) *"If Sagnac and Michelson-Gale, Why Not Michelson-Morley?"* Galilean Electrodynamics, Vol. 1: Nov/Dec.

Hawking, Stephen (2014) *The Beginning of Time.* Lecture: http://www.hawking.org.uk/the-beginning-of-time.html

Henry, Richard Conn (7 July 2005).*The Mental Universe.* Nature 436, 29 | doi:10.1038/436029a.

Herbert, Nick (1985). Notes from Quantum Reality.
http://www.basicincome.com/bp/quantumreality.htm

Hilgevoord, Jan, ed. (1995) *Physics and Our View of the World.*

Hille, Karl (March 23, 2017) *Gravitational Wave Kicks Monster Black Hole Out of Galactic*

Core. https://www.nasa.gov/feature/goddard/2017/gravitational-wave-kicks-monster-black-hole-out-of-galactic-core

Hill, Paul R. (1995) Unconventional Flying Objects: A Former NASA Scientist Explains How UFOs Really Work (Kindle Locations 1261-1265). Hampton Roads Publishing. Kindle Edition.

Holzner, Steven (2004) *Physics II for Dummies*. Hoboken, NJ: Wiley Publishing, Inc.

Holmbberg, Eric (2006) *The Anthropic Principle*. Video produced by BBC Worldwide Ltd.

Hossenfelder, Sabine (2017) *No, physicists have not created "negative mass*. Backreactions blog. http://backreaction.blogspot.com/2017/04/no-physicists-have-not-created-negative.html

Hughes, R.I.G. *The Structure and Interpretation of Quantum Mechanics*. Harvard University Press.

IANDS (International Association for Near Death Studies) (2016) Report of AWARE study. https://iands.org/news/news/front-page-news/1060-aware-study-initial-results-are-published.html

Iowa State Dept. of Physics and Astronomy (2001) *Polaris Project*. http://www.polaris.iastate.edu/EveningStar/Unit2/unit2_sub1.htm.

IPN Progress Report 42-159 2004

Iverson K.E. (1980) "*Notation as A Tool of Thought*", Communications of the ACM, 23: 444–465.

Ives, H. E.; Stilwell, G. R. (1938). "An experimental study of the rate of a moving atomic clock". Journal of the Optical Society of America. 28 (7): 215.

Jaffe, R. (2005). "Casimir effect and the quantum vacuum". Physical Review D. 72 (2): 021301.

Jagerman, Louis (2001) *The Mathematics of Relativity for the Rest of Us*. Victoria, B.C.: Trafford Publishing.

Jansen, K.L.R. (1999) Ketamine (K) and Quantum Psychiatry. Asylum 11 (3) 19-21.

Jorlunde Film Denmark (1985) *Quantum Entanglement Documentary - Atomic Physics and Reality*. Published on youtube in 2014 by Muon Ray.
https://www.youtube.com/watch?v=BFvJOZ51tmc&list=PLtnb8DfCuFNx8x7Jga7K7Wni-eccc3r42

Kaiser, David (2014, Nov. 14). *Is Quantum Entanglement Real?* New York Times Sunday Review: SR10.

Kaku, Michio (2002) Statement on Space as Nothing. Discovery Science Video and (2011) *How the Universe Works*. Video by Discovery Communications produced by Pioneer Productions.

Kaku, Michio (2011) *Michio Kaku Explains String Theory*. Youtube video: https://www.youtube.com/watch?v=kYAdwS5MFjQ

Kaku, Michio (2013) *Is God a Mathematician?* Youtube Video: Big Think Channel https://www.youtube.com/watch?v=jremlZvNDuk.

Kaku, Michio (2015) *Can universes form from "nothing"?* Youtube video: https://www.youtube.com/watch?v=JlcHMI0cC00

Kelly, Alphonsus G. PhD (2005) *Challenging Modern Physics*. Boca Raton, FL: Brown Walker Press.

Kenyon, Dean (2002) Interview in "Unlocking the Mystery of Life." DVD by Illustra Media

Khamehchi, M. A. et al (2017). *Negative-Mass Hydrodynamics in a Spin-Orbit–coupled Bose-Einstein Condensate*, Physical Review Letters DOI: 10.1103/PhysRevLett.118.155301

Kim, Y.H. & Shih, Y. (1999). "Experimental realization of Popper's experiment: violation of the uncertainty principle?" *Foundations of Physics* 29 (12): 1849–1861.

Known Universe (2009) *The Fastest*. Season One: Episode 3. National Geographic Production.

Kolata, Gina (1987) *The Sad Legacy of the Dalkon Shield*. The New York Times: December 6, 1987.

Krasnitz, Michael (2002). *Correlation functions in supersymmetric gauge theories from supergravity fluctuations hHKtions* (PDF). Princeton University Department of Physics: p. 91.

Krauss, Lawrence M. (2012). *A Universe from Nothing: Why There Is Something Rather Than Nothing*. New York: Free Press.

Kuhn, Thomas S. (2012) *The Structure of Scientific Revolutions*. Chicago, IL: University of Chicago Press.

Laureyssens, Dirk (2009).*The Gravitational ETHER of Einstein*. Retrieved from: http://www.mu6.com/einstein.html).

Lee, Penny (1996), "*The Logic and Development of the Linguistic Relativity Principle*", the Whorf Theory Complex: A Critical Reconstruction. John Benjamin's Publishing, p. 84.

Lincoln, Donald (March 13, 2018) *Twin Paradox: the real explanation.* Fermi Lab Youtube site: https://www.youtube.com/watch?v=GgvajuvSpF4

Lincoln, Donald (May 21, 2013) *What is Supersymmetry?* Fermi Lab Youtube site: https://www.youtube.com/watch?v=0CeLRrBAI60

Leplin, Jarrett (1984), *Scientific Realism*, University of California Press.

Lévi-Strauss, C. (1967). *Structural Anthropology*. Translated by Claire Jacobson and Brooke Grundfest Schoepf. New York: Doubleday Anchor Books.

Long, Jeffrey (2010) *Evidence of the Afterlife: The Science of Near-Death Experiences.* New York: HarperCollins.

Maglab (2014) https://nationalmaglab.org/education/magnet-academy/watch-play/interactive/electromagnetic-deflection-in-a-cathode-ray-tube-i

Mandelbaum, Ryan F. (4/21/2017) *No, Scientists Didn't Just Create Negative Mass or Defy the Laws of Physics.* http://gizmodo.com/no-scientists-didnt-just-create-negative-mass-or-defy-1794525465

Markey, Sean (October 8, 2003) *Universe is Finite, "Soccer Ball"-Shaped.* National Geographic News: http://news.nationalgeographic.com/news/2003/10/1008_031008_finiteuniverse.html

Martineau, Harriet (1853) *From the Positive Philosophy of Auguste Comte* (translated London) Vol. I.

Masten, Luke (2006) "Fred Hoyle." In: *The Physics of the Universe.* http://www.physicsoftheuniverse.com/scientists_hoyle.html

Maybury-Lewis, David (1992) *Millennium: Tribal Wisdom and the Modern World.* The Global Television Network: video and book.

Mbarek, Saoussen, Paranjape, M. B. (2014) *Negative mass bubbles in de Sitter space-time.* Phys.

Rev. D 90, 101502(R).

Mike, John (2011) *The Anatomy of a Flying Saucer* (Kindle Location 1907). . Kindle Edition.

Milonni, Peter W. (1994) *The Quantum Vacuum: An Introduction to Quantum Electrodynamics* Academic Press.

Minick, Scot (2002) "Unlocking the Mystery of Life". Video by Illustra Media.

Minutephysics (Sep 13, 2017) *Bell's Theorem: The Quantum Venn Diagram Paradox*. https://www.youtube.com/watch?v=zcqZHYo7ONs&list=PLZ7qfdXfdJe6_0ezTgUZD2d9MBRnFAW8D

Mitchell, William C. (2002) *Bye, Bye Big Bang, Hello Reality*. Carson City, Nevada: Cosmic Sense Books.

Moon, P. and Spencer, D.E. (1956) *"On the Establishment of Universal Time"*, Phil. Sci., Vol. 23, No. 3 (Jul., 1956), pp. 216-229.

Musser, George (2003) *According to the big bang theory, all the matter in the universe erupted from a singularity. Why didn't all this matter--cheek by jowl as it was--immediately collapse into a black hole?* Scientific American: scientificamerican.com/article/according-to-the-big-bang/

Munday, J.N.; Capasso, F.; Parsegian, V.A. (2009). "Measured long-range repulsive Casimir-Lifshitz forces". Nature. 457 (7226): 170–3.

Naeser, C. W. (1979). *Fission-Track Dating and Geologic Annealing of Fission Tracks*. In: Jäger, E. and J. C. Hunziker, Lectures in Isotope Geology, Springer-Verlag.

Nave C. R. (2016) "*Electroweak Unification*". Hyperphysics (Georgia State University). http://hyperphysics.phy-astr.gsu.edu/hbase/Forces/unify.html

Navipedia: *Troposphere Monitoring*. www.navipedia.net

"NAVSTAR GPS User Equipment Introduction" (PDF). US Coast guard navigation center. US Coast Guard. September 1996

Nawrot, W. (1998) "*Some remarks on Around-the-World Atomic Clocks Experiment*", Submitted to International Journal of Theoretical Physics.

Ornstein, Robert (1977) *The Psychology of Consciousness*. New York: Harcourt, Brace, Javanovich.

Ornstein, Robert (1997) *The Right Mind*. New York: Harcourt, Brace, & Company.

Orzel, Chad (2018) *The Real Reasons Quantum Entanglement Doesn't Allow Faster-Than-Light*

Communication. https://www.forbes.com/sites/chadorzel/2016/05/04/the-real-reasons-quantum-entanglement-doesnt-allow-faster-than-light-communication/#699818543a1e

Pandian, Jagadheep D. (June 27, 2015) "Why is the Universe flat and not spherical?" (Advanced)

Ask an Astronomer: http://curious.astro.cornell.edu/about-us/103-the-universe/cosmology-and-the-big-bang/geometry-of-space-time/600-why-is-the-universe-flat-and-not-spherical-advanced

Parapsychological Association (February 11, 2011) *PK on random number generator.* http://www.parapsych.org/articles/36/66/1_pk_on_random_number_generators.aspx

Paul (2016) *Michelson-Morley Experiment Explained.* Youtube video: https://www.youtube.com/watch?v=F2m_VZJM0Zc

Penrose, Roger (1995). *Shadows of the Mind: A Search for the Missing Science of Consciousness.* Oxford: Oxford Univ. Press.

Penrose, Roger (1999). *The Emperor's New Mind: Concerning Computers, Minds, and the Laws of Physics.* Oxford: Oxford Univ. Press.

Petkov, Vesselin (14 December 2003) *Propagation of light in non-inertial reference frames.* Science College, Concordia University. Retrieved from: https://arxiv.org/pdf/gr-qc/9909081.pdf

Pew, Glen (2010). Supersonic Flight, Sonic Booms. AVweb Video: Retrieved from Youtube: https://www.youtube.com/watch?v=gWGLAAYdbbc.

Physics Forums (Oct.19, 2012) https://www.physicsforums.com/threads/do-gravity-waves-imply-repulsive-force-component.645331/

Pike, O, J. et al. (2014) 'A photon–photon collider in a vacuum hohlraum'. *Nature Photonics*, 18 May 2014.

Pogge, Richard (2015) *Real-World Relativity: The GPS Navigation System.* http://www.astronomy.ohio-state.edu/~pogge/Ast162/Unit5/gps.html

Popper, Karl (1962). *Conjectures and Refutations.* New York: Harper Torchbooks.

Popper, K. R. (1967), "Quantum Mechanics Without 'the Observer'", in Mario Bunge (ed.), Quantum Theory and Reality, New York: Springer, pp. 1–12.

Popper, Karl (1982).*Quantum Theory and the Schism in Physics.*

Popper, Karl (1985). "Realism in quantum mechanics and a new version of the EPR experiment" In Tarozzi, G.; van der Merwe, A. Open Questions in Quantum Physics. Dordrecht: Reidel. pp. 3–25.

Powell, Kevin (2016, July/August) *Entanglement.* Discover

Quigg, Chris (2008, January 17). "Sidebar: Solving the Higgs Puzzle". Scientific American.

Reucroft, Stephen and Swain, John (2016) *What is the CASIMIR EFFECT?* Scientific American website, retrieved 12-30-2016, https://www.scientificamerican.com/article/what-is-the-casimir-effec/#checkout

Ricker, Harry H. III *What Happened to Dingle?* http://www.mrelativity.net/Papers/18/Ricker.htm

Sears, Young & Zemansky (1999) *University Physics.* Addison-Wesley. Pp.18–38.

Shankland, R.S. et al. (1955) Rev. Mod. Phys. 27 no. 2, pp. 167–178

Sheldrake, Rupert (2009) *Morphic Resonance.* Rochester: Rock Street Press.

Shifman, Mikhail (October 31, 2012) *Reflections and Impressionistic Portrait at the Conference Frontiers: Beyond the Standard Model*, FTPI (pdf).

Shockey, Peter (2013) *George Rodonaia's – NDE – A Scientist's Afterlife.* Documentary: http://www.lifeafterlife.tv/

Smolin, Lee (2006) *The Trouble with Physics: The Rise of String Theory, The Fall of Science, And What Comes Next.* Harcourt Brace Publishers

Sofaer, A., Zinser, V. & Sinclair, R. M. (1979 a). *A Unique Solar Marking Construct.* Science, 206, 283-291.

Sokal, Alan D. (5 June 1996). "A Physicist Experiments with Cultural Studies". *Lingua Franca.*

Sorensen, Eric (April 17, 2017) *Physicists create 'negative mass.'* https://phys.org/news/2017-04-physicists-negative-mass.htm

Sowell, Thomas (1995) "*Ethnicity and IQ*". In Steven Fraser, ed., *The Bell Curve Wars.* New York: Basic Books, pp. 70-79.

SI Brochure. BIPM (December 22, 2013) Unit of time (second).

Spencer, Domina Eberle & Shama, Uma. *A New Interpretation of the Hafele-Keating Experiment.* http://www.shaping.ru/congress/english/spenser1/spencer1.asp

Stacks Physics Exchange: (2012) http://physics.stackexchange.com/questions/44934/does-matter-with-negative-mass-exist

Styer, Daniel F. (2011) *Relativity for the Questioning Mind.* Baltimore: Johns Hopkins University Press.

Susskind, Leonard (7 July 2008). *The Black Hole War: My Battle with Stephen Hawking to Make the World Safe for Quantum Mechanics.* Hachette Inc.

Susskind, Leonard (25 September 2008). *Quantum Entanglement, Part 1.* Stanford Class Lecture on YouTube: https://www.youtube.com/watch?v=0Eeuqh9QfNI

Susskind, Leonard (Jul 16, 2013) *Why is Time a One-Way Street?* Lecture at Santa Fe Institute. https://www.youtube.com/watch?v=jhnKBKZvb_U

Taylor, John (1980). *Science and the Supernatural: An Investigation of Paranormal Phenomena Including Psychic Healing, Clairvoyance, Telepathy, and Precognition by a Distinguished Physicist and Mathematician.* London: T. Smith.

Talbot, Michael (1992) *The Holographic Universe.* New York: Harper Collins.

Templeton, Graham (November 19, 2012) *Stanford's quantum entanglement device brings us one step closer to quantum cryptography.* Extreme Tech: www.extremetech.com/extreme/140739-stanfords-quantum-entanglement-device-brings-us-one-step-closer-to-quantum-cryptography

Tesla, Nikola (1932) *Space is Nothing.* New York Herald Tribune (September 11)

The Physics Classroom. *The Forbidden F-Word* (1996-2016). http://www.physicsclassroom.com/class/circles/Lesson-1/The-Forbidden-F-Word

Thornbill, Wallace and Talbott, David (2007) *The Electric Universe.* Portland, OR: Mikamar Publishing.

Trubody, Ben (June, 2017) "Richard Feyman's Philosophy of Science." *Philosophy Now*, Issue 120. https://philosophynow.org/issues/114/Richard_Feynmans_Philosophy_of_Science

Tylor, Edward B. (1877) *Primitive Culture.* New York: Henry Holt.

Tytell, David (April 13, 2004) "Building Planets In Plastic-Bags" http://www.skyandtelescope.com/astronomy-news/building-planets-in-plastic-bags/

University of Illinois Physics Website (2007) *Why are atomic masses not expressed as whole numbers?* https://van.physics.illinois.edu/qa/listing.php?id=1216

Veritasium YouTube site (2016) *The Absurdity of Detecting Gravitational Waves.* Interview with Prof Rana Adhikari. https://www.youtube.com/watch?v=iphcyNWFD10&feature=youtu.be

Vessot, R. F. C. and Levine, M. W. (1979, Dec.). Gravitational Redshift Space-Probe Experiment GP-A Project Final Report Contract NAS8-27969 Retrieved from http://ntrs.nasa.gov/archive/nasa/casi.ntrs.nasa.gov/19800011717.pdf.

Von Baeyer, Hans Christian (March 8, 2016) "Quantum Weirdness? It's All in Your Mind" in *Physics at the Limits*, Scientific American Publication.
Readers Respond to "Quantum Weirdness": https://www.scientificamerican.com/article/readers-respond-to-quantum-weirdness/

Wall, Mike (March 24, 2017) *Gravitational Waves Send Supermassive Black Hole Flying*. SPACE.com

Wasson, Valentina and Wasson, Gordon (1957) *Mushrooms, Russia and History*. New York Pantheon, Vol. 2, 264-65.

Weber, Renee (1982) *The Holographic Paradigm* (Fritjof Capra interview) Shambhala/Random House, pp. 217–218.

Wolchover, Natalie (06.30.14) Have We Been Interpreting Quantum Mechanics Wrong This Whole Time? *Quanta Magazines*. https://www.wired.com/2014/06/the-new-quantum-reality/

Wigner, E.P. (1961), "Remarks on the mind-body question", in: I.J. Good, *The Scientist Speculates*, London, Heinemann.

Wikipedia (retrieved 2017) *Faster-than-Light*. https://en.wikipedia.org/wiki/Faster-than-light.

Wikiquotes: https://en.wikiquote.org/wiki/William_Thomson

Woit, Peter (2006). *Not Even Wrong: The Failure of String Theory and the Search for Unity in Physical Law*. Basic Books.

Yaglom, Isaak Moiseevich (1980) Mathematical Structures and Mathematical Modelling. Philadelphia: Gordon and Breach Science Publishers.

White, Leslie (1956) The Locus of Mathematical Reality. In: The World of Mathematics by Newman, James R. New York: Murray Printing Company.

Zato Tomáš (Jun 17 2014) *What-Is-Hawking-Radiation-And-How-Does-It-Cause-A-Black-Hole-To-Evaporate*. Stack Exchange. https://physics.stackexchange.com/questions/26605/what-is-hawking-radiation-and-how-does-it-cause-a-black-hole-to-evaporate

Zukav, Gary. (1979) *The Dancing Wu Li Masters: An Overview of the New Physics*. HarperCollins. Kindle Edition.

www.ingramcontent.com/pod-product-compliance
Lightning Source LLC
Chambersburg PA
CBHW080523240526
45472CB00021BA/1794